Springer Proceedings in Mathematics & Statistics

Volume 104

More information about this series at http://www.springer.com/series/10533

Springer Proceedings in Mathematics & Statistics

This book series features volumes composed of select contributions from workshops and conferences in all areas of current research in mathematics and statistics, including OR and optimization. In addition to an overall evaluation of the interest, scientific quality, and timeliness of each proposal at the hands of the publisher, individual contributions are all refereed to the high quality standards of leading journals in the field. Thus, this series provides the research community with well-edited, authoritative reports on developments in the most exciting areas of mathematical and statistical research today.

Mikhail V. Batsyn • Valery A. Kalyagin
Panos M. Pardalos
Editors

Models, Algorithms and Technologies for Network Analysis

From the Third International Conference on Network Analysis

Editors
Mikhail V. Batsyn
Laboratory of Algorithms and Technologies
  for Networks Analysis
Department of Applied Mathematics
  and Informatics
National Research University Higher
  School of Economics
Nizhny Novgorod, Russia

Valery A. Kalyagin
Laboratory of Algorithms and Technologies
  for Networks Analysis
Department of Applied Mathematics
  and Informatics
National Research University Higher
  School of Economics
Nizhny Novgorod, Russia

Panos M. Pardalos
Department of Industrial
  and Systems Engineering
Center for Applied Optimization
University of Florida
Gainesville, FL, USA

Laboratory of Algorithms and Technologies
  for Networks Analysis
National Research University
  Higher School of Economics
Nizhny Novgorod, Russia

ISSN 2194-1009   ISSN 2194-1017 (electronic)
ISBN 978-3-319-09757-2   ISBN 978-3-319-09758-9 (eBook)
DOI 10.1007/978-3-319-09758-9
Springer Cham Heidelberg New York Dordrecht London

Library of Congress Control Number: 2014949926

Mathematics Subject Classification (2010): 90-02, 90C31, 90C27, 90C09, 90C10, 90C11, 49L20, 90C35, 90B06, 90B10, 90B15, 90B18, 90B40, 90B80, 68R10

© Springer International Publishing Switzerland 2014
This work is subject to copyright. All rights are reserved by the Publisher, whether the whole or part of the material is concerned, specifically the rights of translation, reprinting, reuse of illustrations, recitation, broadcasting, reproduction on microfilms or in any other physical way, and transmission or information storage and retrieval, electronic adaptation, computer software, or by similar or dissimilar methodology now known or hereafter developed. Exempted from this legal reservation are brief excerpts in connection with reviews or scholarly analysis or material supplied specifically for the purpose of being entered and executed on a computer system, for exclusive use by the purchaser of the work. Duplication of this publication or parts thereof is permitted only under the provisions of the Copyright Law of the Publisher's location, in its current version, and permission for use must always be obtained from Springer. Permissions for use may be obtained through RightsLink at the Copyright Clearance Center. Violations are liable to prosecution under the respective Copyright Law.
The use of general descriptive names, registered names, trademarks, service marks, etc. in this publication does not imply, even in the absence of a specific statement, that such names are exempt from the relevant protective laws and regulations and therefore free for general use.
While the advice and information in this book are believed to be true and accurate at the date of publication, neither the authors nor the editors nor the publisher can accept any legal responsibility for any errors or omissions that may be made. The publisher makes no warranty, express or implied, with respect to the material contained herein.

Printed on acid-free paper

Springer is part of Springer Science+Business Media (www.springer.com)

Nothing endures but change.
  Heraclitus (Greek philosopher, c. 535–475 BC)

Preface

Network analysis originated many years ago. In the eighteenth century Euler solved the famous Königsberg bridge problem. Euler's solution is considered to be the first theorem of network analysis and graph theory. In the nineteenth century Gustav Kirchhoff initiated the theory of electrical networks. Kirchhoff was the first person who defined the flow conservation equations, one of the milestones of network flow theory.

After the invention of the telephone by Alexander Graham Bell in the nineteenth century the resulting applications gave the network analysis a great stimulus.

The field evolved dramatically after the nineteenth century. The first graph theory book was written by Dénes König in 1936. As in many other fields, World War II played a crucial role in the development of the field. Many algorithms and techniques were developed to solve logistic problems from the military. After the war, these technological advances were applied successfully in other fields. The earliest linear programming model was developed by Leonid Kantorovich in 1939 during World War II, to plan expenditures to reduce the costs of the army.

In 1940 Tjalling Koopmans also formulated linear optimization models to select shipping routes to send commodities from America to specific destinations in England. For their work in the theory of optimum allocation of resources these two researchers were awarded with the Nobel Prize in Economics in 1975.

The first complete algorithm for solving the transportation problem was proposed by Frank Lauren Hitchcock in 1941. With the development of the Simplex Method for solving linear programs by George B. Dantzig in 1957, a new framework for solving several network problems became available. The network simplex algorithm is still in practice, one of the most efficient algorithms for solving network flow problems. Many other authors proposed efficient algorithms for solving different network problems. Joseph Kruskal in 1956 and Robert C. Prim in 1957 developed algorithms for solving the minimum spanning tree problem. In 1956 Edsger W. Dijkstra developed his algorithm for solving the shortest path problem, one of the most recognized algorithms in network analysis.

As it happened in other fields, computer science and networks influenced each other in many aspects. In 1963 the book by Lester R. Ford and Delbert R. Fulkerson introduced new developments in data structure techniques and computational complexity into the field of networks.

In recent years the evolution of computers has changed the field. We are now able to solve large-scale network problems. In addition, new approaches and computer environments such as parallel computing, grid computing, cloud computing, and quantum computing have helped to solve very large-scale network optimization problems.

Network Analysis has become a major research topic over the last years. The broad range of applications that can be described and analyzed by means of a network is bringing together researches from numerous fields such as Operations Research, Computer Science, Transportation, Biomedicine, Energy, Social Sciences, and Computational Neuroscience. This remarkable diversity of the fields that take advantage of Network Analysis makes the endeavor of gathering up-to-date material a very useful task.

The objective of Net 2013 conference was to initiate joint research among different groups, in particular Center for Applied Optimization (CAO) at the University of Florida and the Laboratory of Algorithms and Technologies for Networks Analysis (LATNA) in Nizhny Novgorod.

We would like to take this opportunity to thank all the contributing authors, the referees for their constructive efforts to improve the quality of the chapters, and CAO and LATNA for support. Research of the editors is partially supported by NRU HSE, RF government grant, ag.11.G34.31.0057.

Nizhny Novgorod, Russia	Mikhail V. Batsyn
Nizhny Novgorod, Russia	Valery A. Kalyagin
Gainesville, FL, USA	Panos M. Pardalos

Contents

A Method of Static and Dynamic Pattern Analysis
of Innovative Development of Russian Regions in the Long Run 1
Fuad Aleskerov, Ludmila Egorova, Leonid Gokhberg, Alexey
Myachin, and Galina Sagieva

Using Mathematical Programming to Refine Heuristic
Solutions for Network Clustering 9
Sonia Cafieri and Pierre Hansen

Market Graph Construction Using the Performance Measure
of Similarity 21
Andrey A. Glotov, Valery A. Kalyagin, Arsenii N. Vizgunov,
and Panos M. Pardalos

The Flatness Theorem for Some Class of Polytopes
and Searching an Integer Point 37
Dmitry V. Gribanov

How Independent Are Stocks in an Independent Set
of a Market Graph 45
Petr A. Koldanov and Ivan Grechikhin

Analysis of Russian Industries in the Stock Market 55
Nina N. Lozgacheva and Alexander P. Koldanov

A Particle Swarm Optimization Algorithm for the Multicast
Routing Problem 69
Yannis Marinakis and Athanasios Migdalas

König Graphs for 4-Paths 93
Dmitry Mokeev

A Hybrid Metaheuristic for Routing on Multicast Networks 105
Carlos A.S. Oliveira and Panos M. Pardalos

**Possible Ways of Applying Citations Network Analysis
to a Scientific Writing Assistant** 119
Alexander Porshnev and Maxim Kazakov

**Bounding Fronts in Multi-Objective Combinatorial
Optimization with Application to Aesthetic Drawing
of Business Process Diagrams**... 127
Julius Žilinskas and Antanas Žilinskas

A Method of Static and Dynamic Pattern Analysis of Innovative Development of Russian Regions in the Long Run

Fuad Aleskerov, Ludmila Egorova, Leonid Gokhberg, Alexey Myachin, and Galina Sagieva

Abstract The term "pattern" refers to a combination of values of some features such that objects with these feature values significantly differ from other objects. This concept is a useful tool for the analysis of behavior of objects in both statics and dynamics. If the panel data describing the functioning of objects in time is available, we can analyze pattern changing behavior of the objects and identify either well adapted to the environment objects or objects with unusual and alarming behavior.

In this paper we apply static and dynamic pattern analysis to the analysis of innovative development of the Russian regions in the long run and obtain a classification of regions by the similarity of the internal structure of these indicators and groups of regions carrying out similar strategies.

1 Introduction

In this paper, the term "pattern" is referred to a combination of values of some features such that these feature values define a group of objects that significantly differ from other objects. The concept of 'pattern' can be defined in three equivalent mathematical frameworks that appeal to different cognitive subsystems pertaining to image, logics, and geometry, respectively. The first approach utilizes parallel coordinates for the visual analysis in order to determine different patterns, the second approach uses conjunctive interval predicates that define a set of classifiers separating the patterns from each other and from the rest, and the third approach represents a pattern as the Cartesian product of the corresponding intervals [11].

The static pattern analysis is a two-step method of the automated pattern formation. At the first stage we use classical cluster analysis to find clusters of objects, and at the second stage we find patterns that adequately represent the obtained clusters. Dynamic analysis of patterns is to highlight the types of functional

---

F. Aleskerov • L. Egorova (✉) • L. Gokhberg • A. Myachin • G. Sagieva
National Research University Higher School of Economics, 20 Myasnitskaya str., Moscow 101000, Russia
e-mail: legorova@hse.ru

stability of objects depending on how frequently an object changes between the patterns to which it belongs at different time periods. This typology may help in managing objects; also, it allows us to determine groups of risk consisting of objects changing too frequently. Objects with a stable pattern over time are of a special interest because they can represent objects well adapted to their environment.

These methods were applied to the analysis of real data in: (a) comparative macroeconomic analysis [2, 5], (b) evaluation of efficiency and analysis of business models attended to by banks in Turkey and Russia [1, 3, 4, 7, 8], (c) patterns of voting behavior and electoral change in General Elections in the UK and Municipal Elections in Finland [6, 9]. Below we use static and dynamic pattern analysis for issues of innovation in the development of regions in the Russian Federation in the long run and investigate the relation between the level of socio-economic conditions and the potential and efficiency of science and education in region and its innovative activity. The successful development of the regions in the long run should be strongly connected with economic and social factors determined by the degree of development of education and science in the region. Any innovative activity is determined by the level of scientific development in the region and is not possible without properly trained personnel, i.e. scientists, students of all levels of education, and qualified employees.

2 Method Description

2.1 Basic System of Indicators

On the first stage we constructed a base system of indicators. It is based on a system of indicators designed for ranking of regional innovative development [13]. That system of indicators is multi-layered and includes 4 groups of indicators, describing social and economic conditions of innovative activity (macroeconomic fundamentals, characteristics of the educational potential of population), scientific and technical potential of innovative activity (including personnel and financial potential, publication, and patent activity), innovative activity (the activity in the field of technological and non-technological innovations, the development of innovative small business, the cost of technological innovation, and the impact of innovation), and the quality of the regional innovative policy (quality of the regulatory framework and organizational support for innovative policy, volume of consolidated budget expenditures in regions).

Based on this system of indicators we have constructed a different feature system, including such sets of indicators, as (1) socio-economic conditions in the region; (2) educational potential of the region's population; (3) potential for research; (4) potential for innovative activity; (5) efficiency of innovative activity. All data used has been taken from the Handbooks of the Federal State Statistics Service [19–21] and the Statistical Handbooks of HSE [14], the data covers 4 years from

2007 to 2010. Initially there was sixth group of indicators named "efficiency of research", but we found a significant correlation between this and third group and decided to use only indicator of research potential of the region. Then as in [13] the normalized values of each indicator for all regions are defined as the ratio of the difference between the value of the indicator in the region and the minimum value of it for all regions divided to the difference between the maximum and minimum values of this indicator for all regions

$$\tilde{z}_i^x = \frac{z_i^x - \min_x z_i^x}{\max_x z_i^x - \min_x z_i^x}$$

where i stands for the number of indicator, x denotes the region, z_i^x denotes the value of indicator i in the region x. So, the normalized indicators are changed from 0 (the region with the minimum value of the indicator) to 1 (the region with the maximum value of the indicator).

We used these indices to characterize the region in the context of the development of science, education, and innovative activity and form a description of the region $(z_1^x, z_2^x, z_3^x, z_4^x, z_5^x)$ in five-dimensional feature space.

2.2 Description of the Pattern Construction

Let X denote a set of objects and Y denote a set of indices (names, labels) of clusters. A distance (metric) $\rho(x, x')$ between objects is specified for a formal description of "closeness" of objects $x, x' \in X$. The set of objects X is split into disjoint subsets, called clusters, so that each cluster consists of objects that are close with respect to chosen metric, and objects of different clusters are significantly different. The cluster number y_i is assigned to each object $x_j \in X$.

The description of the regions we present in a system of parallel coordinates [12, 15], replacing the point $(z_1^x, z_2^x, z_3^x, z_4^x, z_5^x)$ in five-dimensional feature space by piecewise-linear functions. These functions are constructed as follows: on the x axis we put indices that characterize the structure of the objects; the y axis represents the values of these parameters. For each object we have a set of points that correspond to the values of these 5 indices. The piecewise-linear function $f^x(t)$ is constructed by connecting these points by straight lines

$$f^x(t) = k_i^x t + a_i^x, i \leq t \leq i + 1,$$

$$k_i^x i + a_i^x = z_i^x,$$

$$k_i^x(i+1) + a_i^x = z_{i+1}^x, i = 1, \ldots, n - 1$$

An example of a set of obtained piecewise-linear functions is shown in Fig. 1.

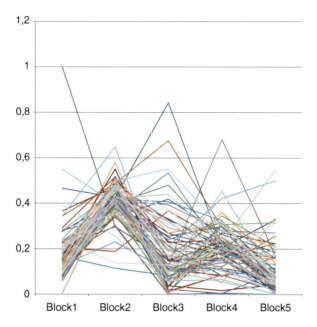

Fig. 1 The piecewise-linear functions that represent 83 regions of Russian Federation, for this example the data was taken only for 2007

In fact, we do not use the absolute values of indices in the pattern analysis and determine patterns not by the position of the points on the axis y but by the slopes of the corresponding lines, i.e., we use the slopes of piecewise-linear functions in clustering. This algorithm produces clusters in which objects from one cluster have same structure, although the absolute values of the features may differ. For example, if objects A and B are characterized by vectors (20, 50, 60) and (2, 5, 6), then these two objects will form two different clusters in the conventional cluster analysis, while the A and B objects obviously have similar structure (although the object A is 10 times bigger than object B). Formally we assign vector $V_i^x = (k_i^x, a_i^x)$ to all intervals $[i, i + 1]$ and all functions f^x can be represented by vectors $V^x = \{V_1^x, \ldots, V_n^x\}$. As a distance between two points V^{x_1} and V^{x_2} the Euclidian distance, the Manhattan distance, etc. can be used for clustering. Different methods of cluster analysis can be used (see [18] for description of cluster analysis methods).

The seeming chaotic conglomeration of lines in Fig. 1 shows some regularities and one can see that many regions may have not equal values of indices, but the structures of them are very similar. It is reasonable to assume that if the regions have similar shapes of the piecewise-linear functions, then they have a similar structure of the indices and, therefore, these regions maintain similar models of science, education, and innovative activity development.

3 Results

Taking into account that our purpose is an analysis of the dynamic behavior of the regions, the resulting piecewise-linear functions for each year 2007–2010 were merged into the total sample. We had $83 * 4 = 332$ objects for clustering. In this paper we have combined two methods of clustering: k-means and hierarchical clustering method [18], in every method Euclidean distance was used. We obtained 24 patterns consisting of more than two objects within the pattern, and 24 patterns with single object in each of them. Some of these patterns are shown in Figs. 2 and 3.

Pattern 1 contains 11 objects representing the city of Moscow and the city of St. Petersburg for all 4 years from 2007 to 2010, and the Primorsky Krai in 2007, 2009, 2010. This pattern is characterized by moderate (0.3–0.5) values of the first two indices ("Socio-economic conditions" and "Educational potential"), high and somewhat extremely high (0.5–0.9) value of "Effectiveness of research", and very low (less than 0.2) values of indices 4 and 5 scoring the potential and effectiveness of innovative activity.

It is interesting to see that (1) Moscow and Saint Petersburg belong to one pattern which means that they have similar internal structure of science, education, and

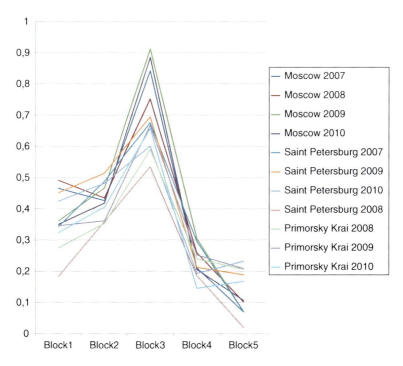

Fig. 2 Pattern 1

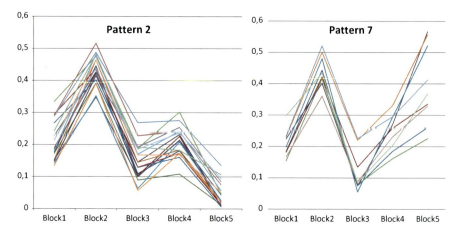

Fig. 3 Patterns 2 and 7

innovation activity, and (2) these cities belong to one pattern over all years which means that they are stable and keep the chosen strategy of development.

In Pattern 2 there are 19 objects. The main characteristics of this pattern are low (0.1–0.3) values for indices 1, 3, 4, 5, and medium (0.4–0.5) values of "Educational potential".

Pattern 7 contains 9 objects with high effectiveness of innovative activity, with moderate educational potential and low level of the remaining indices.

Dynamic pattern analysis was conducted as well, which allows tracking what pattern each of the regions followed 4 years 2007–2010 based on the trajectories of the object. The trajectory of the object is an alternation of patterns, which describes changes of the object on the horizon of the analysis. Examples of such trajectories are shown in Fig. 4.

Pattern analysis allowed us to find tendencies and regularities that we could not identify by classical methods of data mining and cluster analysis. The Russian regions characteristics have been examined in dynamics from 2007 to 2010 and we have obtained a classification of regions with respect to the similarity of the internal structure of science, education, and innovation activity, also we have found groups of regions with high degree of development of these indicators and dynamic classification of regions keeping the chosen strategy of development [10].

Based on the static and dynamic pattern analysis a software AIDA was developed [17]. This software is designed for intelligent mining on large amounts of statistical data in the domains of science, education and innovation activity. The AIDA system provides the following functions:

- identification of trends in the time series of different indicators;
- identification of the implicit mutual correlation between indicators;
- identification of atypical dynamics of indicators;
- assessment of the consistency between different indicators changes;

Static and Dynamic Pattern Analysis of Russian Regions Development in the Long Run

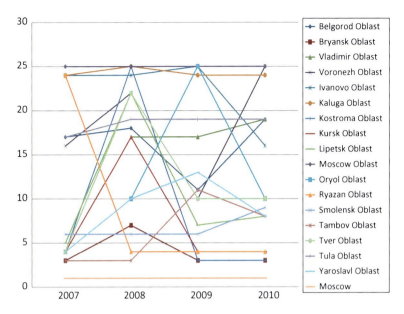

Fig. 4 Trajectories of some regions in Central Federal District over patterns. The numbers of patterns are on the *y* axis

- identification of the mutual correlation of the indicators dynamic trajectories over time;
- determination of specific directions of indicators evolution over time and its qualitative assessment for various analytical measurements;
- determination of the best performance regions and their qualitative assessment.

The AIDA system provides a solution to the complex tasks of statistical information analysis by integration of classical mathematical methods of correlation analysis, pattern analysis, and time series analysis with methods of their semantic interpretation [16]. Further research in this direction should include development of methods and software for forecasting the dynamics of changes of science, education, and innovation activity indicators, for the automated construction of patterns and creation of decision support systems.

Acknowledgements This work is a part of a project of data analysis of science, education, and innovative activity performed by National Research University Higher School of Economics under the state contract No. 07.514.11.4144 "Development of an experimental sample of statistical analysis of science, education, and innovation software using advanced techniques: pattern analysis and data ontological modeling"Ï with Ministry of Education and Science, code 2012-1.4-07-514-0041.

Authors express their sincere gratitude to the Laboratory of Decision Choice and Analysis NRU HSE (F. Aleskerov, L. Egorova, A. Myachin) and Laboratory of Algorithms and Technologies for Network Analysis NRU HSE, Russian Federation Government Grant N. 11.G34.31.0057 (L. Egorova) for partial financial support. The study was undertaken in the framework of the Program of Fundamental Studies of the Higher School of Economics in 2012–2013.

References

1. Aleksashin, P.G., Aleskerov, F.T., Belousova, V.Yu., Popova, E.S., Solodkov, V.M.: Dynamic Analysis of Russian Banks' Business Models in 2006–2009 (in Russian). Working paper WP7/2012/03; Moscow: Publishing House of the University "Higher School of Economics" (2012)
2. Aleskerov, F., Alper, C.E.: Inflation, Money, and Output Growths: Some Observations. Bogazici University Research Paper, SBE, 96–06 (1996)
3. Aleskerov, F., Ersel, H., Gundes, C., Minibas, A., Yolalan, R.: Environmental Grouping of Bank Branches and their Performances. Yapi Kredi Discussion Paper Series, Istanbul, Turkey (1997)
4. Aleskerov, F., Ersel, H., Gundes, C., Yolalan, R.: A Multicriterial Method for Personnel Allocation among Bank Branches. Yapi Kredi Discussion Paper Series, Istanbul, Turkey (1998)
5. Aleskerov, F., Alper, C.E.: A clustering approach to some monetary facts: a long-run analysis of cross-country data. Jpn. Econ. Rev. **51**, 555–567 (2000)
6. Aleskerov, F.: Patterns of Party Competition in British and Russian General and Finnish Municipal Elections. In: Proceedings of the 4-th International Conference on Operational Research, Moscow (2004a)
7. Aleskerov, F., Ersel, H., Yolalan, R.: Multicriterial ranking approach for evaluating bank branch performance. Int. J. Inform. Tech. Decis. Making **3**, 321–335 (2004b)
8. Aleskerov, F.T., Martynova, Y.I., Solodkov, V.M.: Assessment and analysis of the efficiency of banks and banking systems. In: Proceedings of the 2d International Conference "Mathematical Modeling of Social and Economic Dynamic", (MMSED - 2007), Moscow, 13–15 (2007)
9. Aleskerov, F., Nurmi, H.: A method for finding patterns of party support and electoral change: An analysis of British general and finnish municipal elections. Math. Comput. Model. 1225–1253 (2008)
10. Aleskerov, F., Gokhberg, L., Egorova, L., Myachin, A., Sagieva, G.: Study of science, education and innovation data using the pattern analysis(in Russian). Working paper WP7/2012/07, Publishing House of the University "Higher School of Economics", Moscow (2012)
11. Aleskerov, F., Belousova, V., Egorova, L., Mirkin, B.: Methods of pattern analysis in statics and dynamics, part 1: Literature review and classification of the term (in Russian). Bus. Informatics **3**(25), 3–18 (2013)
12. Few, S.: Multivariate Analysis Using Parallel Coordinates http://www.perceptualedge.com/articles/b-eye/parallel_coordinates.pdf. Cited 15 May 2013
13. Gokhberg, et al.: Innovative Development Rating of Regions of Russian Federation (in Russian). Publishing House of the University "Higher School of Economics", Moscow (2012)
14. Indicators of innovative activity: 2008. Statistical Yearbook. http://issek.hse.ru/news/49369377.html. Cited 15 May 2013
15. Inselberg, A.: Parallel Coordinates: Visual Multidimensional Geometry and Its Applications. Springer, New York (2009)
16. Khoroshevsky, V.F.: On a method of semantic interpretation of data patterns based on a structural approach (in Russian). Working paper WP7/2012/08, Publishing House of the University "Higher School of Economics", Moscow (2012)
17. Khoroshevsky, V.F., Moskovskiy, A., Rovbo, M.: Russian regions activity indicators analysis software system based on ontological models and data patterns. Programmnye Produkty i Sistemy **3**, 194–202 (2013)
18. Mirkin, B.: Clustering for Data Mining: A Data Recovery Approach. Chapman and Hall/CRC, Francis and Taylor, Roca Baton (2005)
19. Statistical handbook of the Federal Service of State Statistics "Russian Statistical Yearbook". http://www.gks.ru/bgd/regl/b11_13/Main.htm. Cited 15 May 2013
20. Statistical handbook of the Federal State Statistics Service, "Household Survey on employment". http://www.gks.ru/bgd/regl/b12_30/Main.htm. Cited 15 May 2013
21. Statistical handbook of the Federal State Statistics Service, "Regions of Russia. Socio-economic indicators". http://www.gks.ru/bgd/regl/b11_14p/Main.htm. Cited 15 May 2013

Using Mathematical Programming to Refine Heuristic Solutions for Network Clustering

Sonia Cafieri and Pierre Hansen

Abstract We propose mathematical programming-based approaches to refine graph clustering solutions computed by heuristics. Clustering partitions are refined by applying cluster splitting and a combination of merging and splitting actions. A refinement scheme based on iteratively fixing and releasing integer variables of a mixed-integer quadratic optimization formulation appears to be particularly efficient. Computational experiments show the effectiveness and efficiency of the proposed approaches.

1 Introduction

Networks, or graphs, provide very useful tools for modeling complex systems [33]. They consist of a set V of vertices associated with the entities under study and a set E of edges each of which joins two vertices and corresponds to relationships among the entities. For instance, in sociology vertices are associated with people and edges with relationships like friendship, communication, or collaboration between them. In biology, vertices are associated, for instance, with proteins and the edges with their interactions. Some topological features of networks are studied to better understand the underlying complex systems, as they may reveal the organizational principles of the system components. The structure of complex systems can in fact be understood by identifying the way the nodes of the corresponding networks are connected to each other. A modular structure characterizes many complex systems, meaning that they contain subgroups of entities sharing some common properties. A topic of particular interest in the study of complex networks is therefore the identification of modules, also called *clusters* or *communities*. Given a graph

---

S. Cafieri (✉)
ENAC, MAIAA, F-31055 Toulouse, France and
Université de Toulouse, IMT, F-31400 Toulouse, France
e-mail: sonia.cafieri@enac.fr

P. Hansen
GERAD, HEC Montréal, Canada
e-mail: pierre.hansen@gerad.ca

© Springer International Publishing Switzerland 2014
M.V. Batsyn et al. (eds.), *Models, Algorithms and Technologies for Network Analysis: From the Third International Conference on Network Analysis*, Springer Proceedings in Mathematics & Statistics 104, DOI 10.1007/978-3-319-09758-9__2

$G = (V, E)$, roughly speaking one seeks subgraphs induced by sets of vertices $S_i \subseteq V$ which contain more inner edges (with both vertices in the same subset) than cut edges (with vertices in different subsets). In the last decade the problem of finding clusters in complex networks has been very extensively studied, see Fortunato [15] for an in-depth survey.

Many definitions of network modules have been proposed as well as criteria to evaluate partitions of vertices in modules. Maximizing any such criterion over the set of all partitions is a combinatorial optimization problem. The most popular criterion, despite some recent criticism [8, 16], is the *modularity* of a subnetwork [32]. The modularity of a module is defined as the difference of the fraction of the edges that it contains and the expected number of such edges in a network where edges are distributed at random while keeping the degree distribution of vertices constant, according to the so-called configuration model. Modularity of a partition is the sum of modularities of its clusters. So modularity of a network is a criterion whose maximization provides both the optimal number of clusters and an estimate of the amount of modularity of the network. Numerous heuristics have been proposed for maximizing modularity of a network. They include applications of simulated annealing [20, 28, 29], mean field annealing [26], genetic search [36], extremal optimization [14], variable neighborhood search [3], spectral clustering [31], linear programming followed by randomized rounding [1], dynamical clustering [5], multilevel partitioning [13], contraction–dilation [30], divisive [9, 31] or agglomerative [4, 11] hierarchical clustering, and several other approaches.

Mathematical programming allows us rigorous formulations and solutions for the maximizing modularity optimization problem. Nevertheless, it is rarely used. There are two approaches to use mathematical programming formulations which can be solved to global optimality. Grötschel and Wakabayashi's [18, 19] model for clique partitioning can be immediately applied, replacing the original graph by a complete weighted graph. A closed model is used by Brandes et al. [6]. The second approach was proposed by Xu et al. [39], who express modularity maximization as a mixed-integer quadratic programming problem with a continuous convex relaxation. Column generation can be applied to solve both models [2]. In these models, modularity is the objective function to be maximized and constraints are used to impose conditions defining a partition of the vertex set.

The obtained optimization problems are generally difficult to solve and only small or medium-scale problems can be easily treated. The situation is more favorable when subgraphs of an original graph are handled, as they are more likely to have smaller size (possibly, medium-scale) than the original graph. Given a partition found by a heuristic, one can attempt to refine the result to obtain a new better partition, acting on subnetworks induced by the clusters of the original partition. The purpose of the present paper is to discuss and advance the use of mathematical programming to refine heuristic solutions for network clustering. Two approaches are discussed and compared, one of which is new. The first one was proposed in [10] and is based on splitting clusters using an exact algorithm for bipartitioning and merging pairs of clusters. The new one is inspired by the approach in [38] and is based on iteratively fixing integer variables and solving the corresponding problem.

The paper is structured as follows. In Sect. 2 we describe the proposed mathematical programming-based approaches to refine heuristic partitions. In particular, a mixed-integer quadratic model for modularity-maximizing clustering is recalled and the two strategies to refine partitions, that use such a model, are presented. In Sect. 3 a computational analysis and comparison, on a set of instances from the literature, is presented and discussed. Section 4 concludes the paper.

2 Mathematical Programming-Based Clustering Refinement

Let us consider a partition found by a heuristic for network clustering. It is constituted by subnetworks induced by the clusters found. As a heuristic has been applied, there is no guarantee that the partition given by these subnetworks represents the optimal solution. Thus, one can seek an improved solution by applying a refinement technique.

We propose in this section mathematical programming-based refinement techniques, to be employed as post-processing of heuristics for modularity maximization. First, we recall the main elements of a mixed-integer quadratic model for modularity maximization which is used in these refinement techniques.

2.1 A MIQP Mathematical Programming Model

Let $G = (V, E)$ be an undirected unweighted graph, with set of vertices V of order $n = |V|$ and set of edges E of size $m = |E|$. Modularity Q of G can be expressed as the sum of modularities of clusters, each one being a function of its number of inner edges and of the sum of degrees of its vertices:

$$Q = \sum_s \left[\frac{m_s}{m} - \left(\frac{D_s}{2m} \right)^2 \right], \tag{1}$$

where m_s denotes the number of edges in cluster s, and D_s denotes the sum of degrees k_i of the vertices of cluster s.

In [39] a mixed-integer quadratic formulation is proposed, where (1) is the objective function to be maximized and binary variables are used to identify to which cluster each vertex and each edge belongs. Sets of allocation constraints, and constraints used to express that each vertex belongs to exactly one module, to impose lower and upper bounds on the cardinality of the modules and to break symmetries, fully define the model. In [10] this model is specialized to the case of two clusters only, i.e., a bipartition of the graph. Such a model for bipartitioning is recalled below. Notice that it has been also successfully used to build a hierarchical divisive clustering algorithm, see [7,9].

First observe that in the case of bipartitioning the objective function (1) can be rewritten in a simpler form, expressing the sum of degrees of vertices belonging to one of the two clusters, say D_2, as a function of the sum of degrees D_1 of vertices belonging to the other one: $D_2 = D_c - D_1$, where D_c denotes the sum of degrees in the cluster c to be bipartitioned. The objective function to split cluster c can then be written as the following quadratic function:

$$Q_c = \frac{m_1 + m_2}{m} - \frac{D_1^2}{2m^2} - \frac{D_c^2}{4m^2} + \frac{D_1 D_c}{2m^2}. \quad (2)$$

where m_1 and m_2 are, respectively, the number of edges inside the two clusters.

Decision variables are variables $X_{i,j,s}$ for each edge (v_i, v_j) and $s = 1, 2$, with $X_{i,j,s}$ equal to 1 if the edge (v_i, v_j) is inside cluster s and 0 otherwise, and variables $Y_{i,1}$ for $i = 1, 2, \ldots n$, equal to 1 if the vertex v_i is inside cluster 1 and 0 otherwise. Constraints on the problem are allocation constraints, used to impose that any edge (v_i, v_j) can belong to cluster s if and only if both of its end vertices i and j also belong to that cluster:

$$\forall (v_i, v_j) \in E_c \quad X_{i,j,1} \leq Y_{i,1} \quad (3)$$

$$\forall (v_i, v_j) \in E_c \quad X_{i,j,1} \leq Y_{j,1} \quad (4)$$

$$\forall (v_i, v_j) \in E_c \quad X_{i,j,2} \leq 1 - Y_{i,1} \quad (5)$$

$$\forall (v_i, v_j) \in E_c \quad X_{i,j,2} \leq 1 - Y_{j,1} \quad (6)$$

Further constraints express the number of edges of each of the two clusters and the sum of vertex degrees of the first cluster in terms of the decision variables X and Y, and finally integrality constraints are imposed on variables Y. Notice that integrality of variables X is implied by constraints (3)–(6), as well as integrality of D_1 follows by its defining constraint. The following mixed-integer quadratic (MIQP) model, that has a continuous convex relaxation, is finally obtained [10]:

$$(\mathcal{B}) \begin{cases} \max & Q_c \\ \text{s.t.} & \forall (v_i, v_j) \in E_c \quad X_{i,j,1} \leq Y_{i,1} \\ & \forall (v_i, v_j) \in E_c \quad X_{i,j,1} \leq Y_{j,1} \\ & \forall (v_i, v_j) \in E_c \quad X_{i,j,2} \leq 1 - Y_{i,1} \\ & \forall (v_i, v_j) \in E_c \quad X_{i,j,2} \leq 1 - Y_{j,1} \\ & \forall s \in \{1, 2\} \quad m_s = \sum_{(v_i, v_j) \in E_c} X_{i,j,s} \\ & \quad D_1 = \sum_{v_i \in V_c} k_i Y_{i,1} \\ & \forall s \in \{1, 2\} \quad m_s \in \mathbb{R} \\ & D_1 \in \mathbb{R} \\ & \forall v_i \in V_c \quad Y_{i,1} \in \{0, 1\} \\ & \forall (v_i, v_j) \in E_c \quad \forall s \in \{1, 2\} \quad X_{i,j,s} \in \mathbb{R}_0^+. \end{cases}$$

2.2 Splitting and Merging Clusters

In [10] we proposed a refinement technique for clustering results that is built on the mathematical programming formulation ($\mathscr{B}$) recalled above. First, clusters are considered one at a time and the bipartitioning problem ($\mathscr{B}$) is solved exactly, then pairs of clusters are merged and the exact bipartitioning is applied again. More precisely, in a sequence of steps, starting from the original partition obtained applying a heuristic, each cluster is first bipartitioned using an exact algorithm. Notice that ($\mathscr{B}$) is a MIQP with a continuous convex relaxation, that can be solved to global optimality by any standard solver for MIQP problems through the standard branch-and-bound method. If the modularity value corresponding to the obtained bipartition is higher than the one of the original cluster, then such original cluster is replaced by the new ones obtained by bipartition, otherwise the original cluster is kept. This sequence of bipartition attempts leads to a new, refined partition.

This new partition is further refined by a new sequence of steps, where pairs of clusters, sorted by decreasing number of joining links, are provisionally merged and modularity of the merged cluster is compared to the sum of modularities of the two original clusters. In the case of improvement of the objective function value, the merged cluster is kept at the place of the two original ones. When merging is not beneficial in terms of improvement of the solution, the merged cluster is attempted to be split into two parts, according to the procedure applied in the first sequence of refining steps, exactly solving the bipartition problem. The two new clusters are possibly different from the original ones that have been merged, and can potentially correspond to an improved solution.

2.3 Fixing Integer Variables

We now present a novel mathematical programming-based approach to refine heuristic partitions. It is inspired by the methodology proposed by Xu et al. [38] for community detection in networks. In [38], the authors propose a two-stage procedure, where first a mixed-integer nonlinear problem (similar to that of [39] for a number of clusters generally greater than two, but where the only decision variables are binary variables Y expressing allocation of vertices to modules) is approximately solved to get an initial partition, and then a fixing and releasing scheme is applied. In this second stage, the authors consider the MIQP model in [39] and solve it, by standard solvers, iteratively fixing a certain number of variables Y to their value 1 and releasing the other variables, that are so free to take a value 1 or 0 depending on the way vertices are re-allocated in the current solution. Fixing integer variables gives a mathematical programming formulation with a reduced number of variables, and so more tractable.

We build upon the idea of fixing binary variables, though developing a different approach. Our approach is devised to refine approximate clustering solutions, so we start from the partition provided by a clustering heuristic, that replaces the first stage of the procedure in [38]. Then, we attempt to improve the original partition by acting on modules through a new heuristic based on variable fixing. Starting from an assignment of vertices to modules, i.e., from an assignment of 0–1 values to variables Y, we fix $nfix$ variables to their value 1 and compute a new value for the remaining variables, that is, we re-allocate the corresponding vertices. For each cluster, the vertices that are reallocated are chosen on the basis of their inner degree (the number of neighbors of a vertex inside the cluster), moving first vertices that have a small inner degree and so are likely to have more connections inside a different cluster than the one they are assigned. A given number of (outer) iterations is performed, each one acting on a set *Fix*, containing variables whose value has to be fixed, and a set *Unfix*, containing variables to be released. To avoid using the same sets *Fix-Unfix* in successive iterations, random perturbations are applied to these sets.

As acting on the whole graph requires to solve a mixed-integer nonlinear problem that may be quite large even with a number of variables that are fixed, and splitting and merging clusters appears to be an effective strategy for refinements [10], we integrate our fixing variables-based strategy in the procedure above based on splitting and merging clusters. To refine a given partition, again we implement the two consecutive steps performing, respectively, bipartitioning of each cluster and merging mixed to bipartitioning on pairs of clusters. Thus, we consider the MIQP formulation ($\mathcal{B}$), but in place of solving exactly the bipartitioning problem by standard branch-and-bound for MIQP, we apply our fixing variables-based strategy.

Thus, our refinement procedure works as follows.

First, each cluster of the original partition is split into two sets. To that effect, an initial approximate solution for the bipartition is computed and the above fixing variables-based approach is applied. If the modularity value corresponding to the obtained bipartition is higher than the one of the original cluster, then the original cluster is replaced by the new ones obtained by bipartition, otherwise the original cluster is kept. Once all clusters of the original partition have been examined, the merging-and-splitting procedure is applied. Pairs of clusters, sorted by decreasing number of joining links, are provisionally merged. If merging improves the objective function value, then the merged cluster is kept, otherwise it is split into two subsets again applying the fixing variables-based approach.

3 Computational Results

In this section, we apply the proposed clustering refinement techniques to the partitions found by two known heuristics for graph modularity maximization. The first one was proposed by Noack and Rotta [34] and is based on a single-step coarsening with a multi-level refinement. The second one was proposed in 2011 by Cafieri et al. [9] and is a hierarchical divisive heuristic that is locally optimal in the sense that bipartitions are computed by an exact optimization algorithm.

The first refinement technique (Sect. 2.2) is implemented solving the mixed-integer quadratic bipartition problem ($\mathscr{B}$) using CPLEX 12.2 [22], setting its parameters in such a way that the MIP cutting plane generation is disabled, the branching variable selection strategy is based on reduced pseudo costs, the number of nodes in the Branch-and-Bound tree is limited to 40000, and 1 only thread is used.

The fixing variables-based technique (Sect. 2.3) is implemented using as a starting guess an (approximate) affectation of variables provided by CPLEX 12.2 limited to the solution at the root node, and then iterating the fixing variables scheme over 100 iterations. At each iteration, the number of fixed variables is set to half the cardinality of the current subgraph.

We test the proposed refinement algorithms on datasets in the literature, which correspond to networks modeling various real-life applications. Specifically, we consider a social network of dolphins [27], a network describing interactions among the characters of Hugo's novel *Les Misérables* [23], a biological network of protein–protein interactions [12], a network recording co-purchasing of political books on Amazon.com [24], a representation of the schedule of games between American college football teams in the Fall of 2000 [17], a network of connections between US airports [35], a network describing electronic circuits [25], e-mail interchanges between members of a university [21], a network giving the topology of the Western States Power Grid of the United States [37], and authors collaborations [35]. The considered datasets are listed in Table 1 together with their number of vertices n and number of edges m. Solutions have been obtained on a 2.4 GHz Intel Xeon CPU of a computer with 8GB RAM shared by three other similar CPU running Linux.

In Tables 2 and 3 we report the results of the refinements of clustering results obtained using the Noack and Rotta's [34] (NR) heuristic and the

Table 1 Datasets in the literature, with their number of vertices n and number of edges m

Dataset	n	m
Dolphins	62	159
Les miserables	77	254
p53_protein	104	226
Political books	105	441
Football	115	613
usair97	332	2126
netscience_main	379	914
s838	512	819
Email	1133	5452
Power	4941	6594
erdos02	6927	11850

Table 2 Modularity values corresponding to the partition found by the Noack and Rotta's heuristic [34] (Q_{NR}), by our first approach for refinement after the splitting step only (Q_{split}^{NR}) and after the merging and splitting step ($Q_{mrg+spl}^{NR}$), and by our fixing variables-based approach after the splitting step only ($Q_{split\_fix}^{NR}$) and after the merging and splitting step ($Q_{mrg+spl\_fix}^{NR}$). In the last column, the optimal modularity value Q_{opt} is reported, when available in the literature [2]

Dataset	Q_{NR}	Q_{split}^{NR}	$Q_{mrg+spl}^{NR}$	$Q_{split\_fix}^{NR}$	$Q_{mrg+spl\_fix}^{NR}$	Q_{opt} [2]
Dolphins	0.52377	0.52773	0.52852	0.52508	0.52646	0.52852
Les miserables	0.56001	0.56001	0.56001	0.56001	0.56001	0.56001
p53_protein	0.53216	0.53216	0.53502	0.53216	0.53502	0.53513
Political books	0.52694	0.52724	0.52724	0.52694	0.52694	0.52724
Football	0.60028	0.60237	0.60457	0.60237	0.60457	0.60457
usair97	0.36577	0.36577	0.36808	0.36577	0.36808	0.3682
netscience_main	0.84745	0.84828	0.84842	0.84828	0.84842	0.8486
s838	0.81624	0.81624	0.81656	0.81624	0.81656	0.8194
Email	0.57740	0.57741	0.57776	0.57741	0.57768	–
Power	0.93854	0.93867	0.93873	0.93854	0.93858	–
erdos02	0.75926	0.75926	0.76958	0.75926	0.78952	–

Table 3 Modularity values corresponding to the partition found by the Cafieri et al.'s heuristic [9] (Q_{CHL}), by our first approach for refinement after the splitting step only (Q_{split}^{CHL}) and after the merging and splitting step ($Q_{mrg+spl}^{CHL}$), and by our fixing variables-based approach after the splitting step only ($Q_{split\_fix}^{CHL}$) and after the merging and splitting step ($Q_{mrg+spl\_fix}^{CHL}$). In the last column, the optimal modularity value Q_{opt} is reported, when available in the literature [2]

Dataset	Q_{CHL}	Q_{split}^{CHL}	$Q_{mrg+spl}^{CHL}$	$Q_{split\_fix}^{CHL}$	$Q_{mrg+spl\_fix}^{CHL}$	Q_{opt} [2]
Dolphins	0.52646	0.52646	0.52680	0.52646	0.52680	0.52852
Les miserables	0.54676	0.54676	0.55351	0.54676	0.55351	0.56001
p53_protein	0.53000	0.53000	0.53004	0.53000	0.53145	0.53513
Political books	0.52629	0.52629	0.52678	0.52629	0.52678	0.52724
Football	0.60091	0.60091	0.60112	0.60091	0.60112	0.60457
usair97	0.35959	0.35959	0.35975	0.35959	0.35960	0.3682
netscience_main	0.84702	0.84702	0.84703	0.84702	0.84703	0.8486
s838	0.81663	0.81663	0.81675	0.81663	0.81667	0.8194
Email	–	–	–			–
Power	0.93937	0.93937	0.93941	0.93937	0.93941	–
erdos02	–	–	–			

Cafieri et al.'s [9] (*CHL*) heuristic, respectively. We compare the results of the mathematical programming-based refinements described in Sect. 2, showing the original modularity value computed by the heuristic under consideration (*NR* or *CHL*), the intermediate result obtained by cluster splitting only and the final result after sequentially applying the splitting step and the merging step mixed to splitting, for the first refinement technique (*split* and *mrg+spl*) (also in [10]) and, respectively, the new one based on fixing variables (*split_fix* and *mrg + spl_fix*). We are able to

Using Mathematical Programming to Refine Heuristic Solutions for Network Clustering   17

Table 4 Computing time (seconds) required by the proposed approaches applied as post-processing to Noack and Rotta's heuristic ($time^{NR}$) and Cafieri et al.'s heuristic ($time^{CHL}$). Solutions have been obtained on a 2.4 GHz Intel Xeon CPU of a computer with 8GB RAM shared by three other similar CPU running Linux

Dataset	$time^{NR}_{mrg+spl}$	$time^{NR}_{mrg+spl\_fix}$	$time^{CHL}_{mrg+spl}$	$time^{CHL}_{mrg+spl\_fix}$
Dolphins	0.20	0.39	0.26	0.20
Les miserables	0.67	0.71	0.35	0.30
p53_protein	1.02	1.23	0.26	0.49
Political books	5.10	1.66	3.41	1.21
Football	3.26	3.16	0.99	0.83
usair97	334.72	8.96	454.64	16.86
netscience_main	1.38	1.67	0.77	0.85
s838	1.20	1.40	1.06	1.16
Email	57.80	56.02	–	–
Power	18.62	15.81	17.50	15.42
erdos02	919.74	241.29	–	–

obtain improved results for all the tested cases out of one (political books) refined with the fixing variables technique. Comparing the refined results with optimal modularity maximization solutions, when available in the literature [2], we remark that in some cases we get the optimal partitions, and in general very good quality solutions. The results obtained applying the two proposed refinements are generally comparable, and often we get the same modularity value (up to 5 decimal digits) in the two cases. When this is not the case, the values coincide up to 2 or 3 decimal digits.

In Table 4 we compare the two proposed approaches in terms of computing time. Very short times are spent in both cases on small-scale networks. For larger networks, it appears that the proposed approach based on fixing integer variables reduces sometimes significantly the time needed to refine the initial partition. This happens, as expected, especially for networks for which exact bipartitioning takes time because of the exploration of a large Branch-and-Bound tree. For example, improving the NR heuristic, time is reduced from 334.72 to 8.96 seconds for the 6-th dataset and from 919.74 to 241.29 seconds for the last dataset, and, improving the CHL heuristic, the reduction is from 454.64 to 16.86 seconds, again for the 6-th dataset.

Figure 1 illustrates the clustering of a network for which the optimal modularity-maximizing partition is obtained refining the NR heuristic result.

4 Conclusions

We proposed mathematical programming-based approaches to refine graph clustering solutions. In particular we discussed and compared two approaches, the one in [10] based on splitting clusters and a combination of merging and splitting

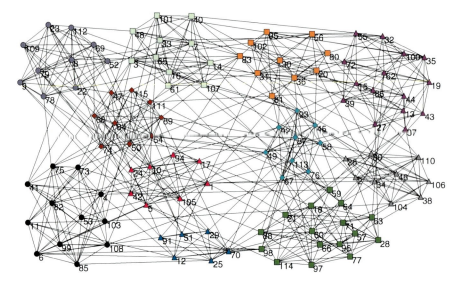

Fig. 1 Optimal clustering of network football obtained refining the *NR* heuristic result

clusters, where bipartitions are computed exactly solving a MIQP problem, and a new one, based on iteratively fixing and releasing integer variables, again integrated in a splitting and merging–splitting scheme. We employ our approach as post-processing of some known heuristics for modularity maximization, obtaining improved solutions and, for some datasets, the optimal partition. The proposed approach based on fixing integer variables allows us to significantly reduce the computing time needed to provide an improved clustering solution.

Acknowledgements The first author has been supported by French National Research Agency (ANR) through grant ANR 12-JS02-009-01 "ATOMIC."

References

1. Agarwal, G., Kempe, D.: Modularity-maximizing graph communities via mathematical programming. Eur. Phys. J. B **66**(3), 409–418 (2008)
2. Aloise, D., Cafieri, S., Caporossi, G., Hansen, P., Liberti, L., Perron, S.: Column generation algorithms for exact modularity maximization in networks. Phys. Rev. E **82**(4), 046,112 (2010)
3. Aloise, D., Caporossi, G., Hansen, P., Liberti, L., Perron, S., Ruiz, M.: Graph partitioning and graph clustering. In: Proceedings of the 10th DIMACS implementation challenge workshop, Atlanta, GA, USA, 2012, Contemporary Mathematics, 588, pp. 113–127. American Mathematical Society (AMS), Oxford (2013)
4. Blondel, V., Guillaume, J.L., Lambiotte, R., Lefebvre, E.: Fast unfolding of communities in large networks. J. Stat. Mech. Theory Exp. P10008 (2008)
5. Boccaletti, S., Ivanchenko, M., Latora, V., Pluchino, A., Rapisarda, A.: Detecting complex network modularity by dynamical clustering. Phys. Rev. E **75**, 045,102 (2007)

Using Mathematical Programming to Refine Heuristic Solutions for Network Clustering

6. Brandes, U., Delling, D., Gaertler, M., Görke, R., Hoefer, M., Nikoloski, Z., Wagner, D.: On modularity clustering. IEEE Trans. Knowledge Data Eng. **20**(2), 172–188 (2008)
7. Cafieri, S., Costa, A., Hansen, P.: Reformulation of a model for hierarchical divisive graph modularity maximization. Ann. Oper. Res. (2012). DOI 10.1007/s10479-012-1286-z (in press)
8. Cafieri, S., Hansen, P., Liberti, L.: Loops and multiple edges in modularity maximization of networks. Phys. Rev. E **81**(4), 046,102 (2010)
9. Cafieri, S., Hansen, P., Liberti, L.: Locally optimal heuristic for modularity maximization of networks. Phys. Rev. E **83**(5), 056,105 (2011)
10. Cafieri, S., Hansen, P., Liberti, L.: Improving heuristics for network modularity maximization using an exact algorithm. Discrete Appl. Math. **163**(1), 65–72 (2014)
11. Clauset, A., Newman, M., Moore, C.: Finding community structure in very large networks. Phys. Rev. E **70**, 066,111 (2004)
12. Dartnell, L., Simeonidis, E., Hubank, M., Tsoka, S., Bogle, I., Papageorgiou, L.: Self-similar community structure in a network of human interactions. FEBS Lett. **579**, 3037–3042 (2005)
13. Djidjev, H.: A scalable multilevel algorithm for graph clustering and community structure detection. Lect. Note Comput. Sci. **4936**, 117–128 (2008)
14. Duch, J., Arenas, A.: Community identification using extremal optimization. Phys. Rev. E **72**(2), 027,104 (2005)
15. Fortunato, S.: Community detection in graphs. Phys. Rep. **486**(3-5), 75–174 (2010)
16. Fortunato, S., Barthelemy, M.: Resolution limit in community detection. Proc. Natl. Acad. Sci. USA **104**(1), 36–41 (2007)
17. Girvan, M., Newman, M.: Community structure in social and biological networks. Proc. Natl. Acad. Sci. USA **99**(12), 7821–7826 (2002)
18. Grötschel, M., Wakabayashi, Y.: A cutting plane algorithm for a clustering problem. Math. Programm. **45**, 59–96 (1989)
19. Grötschel, M., Wakabayashi, Y.: Facets of the clique partitioning polytope. Math. Programm. **47**, 367–387 (1990)
20. Guimerà, R., Amaral, A.: Functional cartography of complex metabolic networks. Nature **433**, 895–900 (2005)
21. Guimerà, R., Danon, L., Diaz-Guilera, A., Giralt, F., Arenas, A.: Self-similar community structure in a network of human interactions. Phys. Rev. E **68**, 065,103 (2003)
22. IBM: ILOG CPLEX 12.2 User's Manual. IBM (2010)
23. Knuth, D.: The Stanford GraphBase: A Platform for Combinatorial Computing. Addison-Wesley, Reading (1993)
24. Krebs, V.: http://www.orgnet.com/ (unpublished)
25. Lab, U.A.: http://www.weizmann.ac.il/mcb/UriAlon/
26. Lehmann, S., Hansen, L.: Deterministic modularity optimization. Eur. Phys. J. B **60**, 83–88 (2007)
27. Lusseau, D., Schneider, K., Boisseau, O., Haase, P., Slooten, E., Dawson, S.: The bottlenose dolphin community of doubtful sound features a large proportion of long-lasting associations. can geographic isolation explain this unique trait? Behav. Ecol. Sociobiol. **54**(4), 396–405 (2003)
28. Massen, C., Doye, J.: Identifying communities within energy landscapes. Phys. Rev. E **71**, 046,101 (2005)
29. Medus, A., Acuna, G., Dorso, C.: Detection of community structures in networks via global optimization. Phys. A **358**, 593–604 (2005)
30. Mei, J., He, S., Shi, G., Wang, Z., Li, W.: Revealing network communities through modularity maximization by a contraction-dilation method. New J. Phys. **11**, 043,025 (2009)
31. Newman, M.: Modularity and community structure in networks. Proc. Natl. Acad. Sci. USA **103**(23), 8577–8582 (2006)
32. Newman, M., Girvan, M.: Finding and evaluating community structure in networks. Phys. Rev. E **69**, 026,113 (2004)
33. Newman, M.E.J.: Networks: An Introduction. Oxford University Press, Oxford (2010)

34. Noack, A., Rotta, R.: Multi-level algorithms for modularity clustering. Lect. Note Comput. Sci. **5526**, 257–268 (2009)
35. http://vlado.fmf.uni-lj.si/pub/networks/data/
36. Tasgin, M., Herdagdelen, A., Bingol, H.: Community detection in complex networks using genetic algorithms. arXiv:0711.0491 (2007)
37. Watts, D., Strogatz, S.: Collective dynamics of 'small-world' networks. Nature **393**(6684), 409–410 (1998)
38. Xu, G., Bennett, Papageorgiou, L., Tsoka, S.: Module detection in complex networks using integer optimisation. Algorithm Mol. Biol. **5**(36) (2010). DOI:10.1186/1748-7188-5-36
39. Xu, G., Tsoka, S., Papageorgiou, L.: Finding community structures in complex networks using mixed integer optimization. Eur. Phys. J. B **60**, 231–239 (2007)

Market Graph Construction Using the Performance Measure of Similarity

Andrey A. Glotov, Valery A. Kalyagin, Arsenii N. Vizgunov, and Panos M. Pardalos

Abstract The paper presents the description of the modification of the stock market graph model. Authors suggest a new similarity measure between stocks. Following the market graph model a vertex represents a stock. For each pair of stocks we calculate a number of weeks when both stocks were profitable simultaneously. This number is used as a weight of edge between two vertices. The structural properties of the graph constructed by means of the suggested measure can be used to evaluate potential profitability, performance of the stock market. The model was applied to Russian stock market and to US stock market for the time period from 2001 to 2011. The main result of the analysis is the description of the peculiarity of the Russian stock market comparing with the well-developed US stock market and observation that constructed US stock market graph follows the power law model.

1 Introduction

The amount of data generated by the stock markets of the different countries is enormous nowadays. The mathematical models can be used to summarize and visualize stock markets data as well as finding the interesting dependencies between

---

A.A. Glotov • A.N. Vizugnov (✉)
National Research University Higher School of Economics, 25/12 Bolshaja Pecherskaja Ulitsa, Nizhny Novgorod, 603155, Russia
e-mail: aaglotov@edu.hse.ru; anvizgunov@hse.ru

V.A. Kalyagin
Laboratory of Algorithms and Technologies for Networks Analysis, Department of Applied Mathematics and Informatics, National Research University Higher School of Economics, Nizhny Novgorod, 603155, Russia
e-mail: vkalyagin@hse.ru

P.M. Pardalos
Department of Industrial and Systems Engineering, Center for Applied Optimization, University of Florida, Gainesville, FL, 32608, USA

Laboratory of Algorithms and Technologies for Networks Analysis, National Research University, Higher School of Economics, Nizhny Novgorod, 603155, Russia
e-mail: pardalos@ise.ufl.edu

data. One of the widely used models of the stock market is the market graph model introduced by Boginski et al. [2]. In the market graph each vertex represents a stock. Two vertices are connected by an edge if the corresponding similarity measure value is larger than or equal to the specified threshold $\theta \in [-1, 1]$. The similarity measure can differ [1] but the most popular measure of the similarity is the pair-wise Pearson correlation between stock returns [2, 5, 6]. The dynamics of the characteristics of the constructed market graph can be used to obtain deeper understanding of the interdependencies between stock returns. At the same time it is possible to obtain the interesting results using other similarity measures between two stocks not only the Pearson correlation.

In our work we change the market graph model by applying a new similarity measure between two stocks. We suggest calculating the number of periods when the both stocks are profitable simultaneously taking into consideration the inflation rate. This number is used as a weight of the edge between two vertices. The higher value of the considered characteristic means more periods of profitability of the stocks. So the market graph model with the suggested similarity measure can be used to evaluate the potential profitability, performance of the market.

The main advantage of the suggested performance measure is the possibility to interpret the results from the economics point of view. It makes the suggested measure more suitable for the analysis of the stock market than the Pearson [2, 5, 6] and sign correlations [1]. It makes also possible to check the obtaining results of the analysis. The most used characteristic describing the stock market profitability in a short and concise form is a stock index. Because both characteristics describe the profitability of the market they should be related. Our calculations show that the behavior of the performance measure corresponds to the index change during the most of the periods.

We consider the following characteristics of the stock market graph: edge weights distribution histogram, edge density and clustering coefficient, following the power law model. We constructed the market graphs with suggested similarity measure for Russian and US stock markets for the time period from 2001 to 2011. The paper is organized as follows. In Sect. 2 we briefly recall the market graph model and describe the modification of the model. In Sect. 3 we describe the Russian and US markets data. In Sect. 4 we present the results of the analysis of the Russian market graph with suggested similarity measure and compare their characteristics with corresponding US market graph. In Sect. 5 we give concluding remarks.

2 Stock Market Graph Models and Their Modification

The market graph model was introduced by Boginski et al. [2]. Each vertex of the market graph model represents a stock. The Pearson correlation of the stock returns is used as a measure of the similarity of the stocks. The edge exists between the vertices if the corresponding correlation coefficient is equal to or greater than the specified threshold. The market graph model is widely used. In particular it was used to analyze US [2–4], Chinese [5], Swedish [1, 6], Russian stock markets [1, 7, 8].

Market Graph Construction Using the Performance Measure of Similarity

Despite the popularity of the Pearson correlation as a measure of the similarity between stocks there are the cases that demand the use of a different measure. For example, in paper [1] authors introduced a new measure for the construction of the market graph referred to as a sign correlation. This measure is based on the probability of the coincidence of the signs of the stock returns. The authors showed that the sign correlation measure is robust, has a simple interpretation, and in some cases allows obtaining better results of the analysis.

In order to evaluate the potential profitability of the stock market we suggest using a new similarity measure instead of the Pearson correlation. For each week of the considered time period we calculate the return of every stock and compare this number with the inflation value. If the return of the stock is higher than the inflation, then we define this stock as profitable during this week. Then for each pair of stocks we calculate a number of weeks when both stocks were profitable simultaneously. This number is used as a weight of an edge between two vertices. More precisely, let $x_i \in \{0, 1\}$ is the indicator that the price of stock x on week i, $i = 1, \ldots, N$, is profitable. Then

$$c_{xy} = \sum_{i=1}^{N} x_i y_i \qquad (1)$$

defines the performance measure value for the stocks x and y over the period of N weeks.

The suggested measure of the similarity has following simple properties:

1. The similarity measure can take a fixed set of values varying from 0 to the number of weeks in the examination period.
2. If we calculate for each stock the number of weeks when it is profitable, then the weight of the edge between two stocks is limited to the minimum for these numbers.
3. If for two stocks there are no weeks, when both of them are profitable, then the similarity measure equals 0. However the opposite proposition is not true.
4. If we use other time periods than weeks, for example months or days, then the value of the measure will differ. The difference will not depend proportionally on the duration change, because the monthly profitability is not related directly to the profitability of the weeks of this month.
5. In order to calculate the measure value of a set of consequent not intersected periods we need to find the sum of the measure values for each of the periods.
6. The transitivity property of the measure does not take place in general but it takes place in some artificial case when all stocks are profitable at the same weeks.

Properties 1–3 follow from the definition of the measure. Property 4 is related to the possibility to choose different basic time period instead of a week. The choosing of the week as a basic time period is justified with the need to have a tradeoff between the calculations of the inflation that is performed monthly in most cases and daily stock prices. It also allows us to reduce the influence of the volatility of the stock markets on the model. Property 5 gives us the way to calculate the measure for a different time periods. Using property 6 we can evaluate the closeness of the real stock market to the artificial case when all stocks are profitable or not profitable simultaneously.

The main advantages of the modified market graph model are taking the inflation into consideration, using exact values instead of sample correlations and the possibility to interpret the results of the analysis from the economics point of view. It is important to have the information adjusted to the inflation rates, as if the inflation is high then the calculated positive return can be a loss in reality. Using exact values instead of sample correlations provides the possibility to calculate the measure for any number of stocks for any time periods. From the economics point of view the performance measure describes the potential profitability of the stock market. So the results obtaining by analysis of the market graph constructed by using the performance measure can be checked by comparison with the results of the other types of stock market analysis. Another advantage of using the performance measure is the possibility to take into account stocks with low liquidity. If the stock is not traded during the particular week it means that this stock is not profitable during this week. The absence of profitability in this case is caused by the lack of liquidity not by the comparison with the inflation rate. This feature is important for the Russian stock market because many stocks have a low liquidity.

3 Data of the Russian and US Stock Markets

In order to analyze the dynamics of the changes in structural properties of the market graph over time we consider eleven time periods from September 1, 2007 to December 18, 2011. The duration of each period is 52 weeks. We examine the stocks traded on Moscow Interbank Currency Exchange Stock Exchange (MICEX SE), NYSE, AMEX, and NASDAQ. Dates and number of the taken stocks of Russian and US stock markets corresponding to each period are summarized in Table 1.

For Russian stock market data we apply a cleaning procedure. We need such procedure to eliminate stocks with very low liquidity. We choose the stocks with the following characteristic: the number of the weeks on which a stock has been transacted exceeds 20 weeks out of the 52 weeks under consideration. If the stock is not traded during the particular week, it means that this stock is not profitable during this week, hence we do not need to adjust the data regarding the weeks without trading transactions. We would like to note that the number of considered stocks for periods 1–5 for Russian stock market is too small so it is difficult to obtain meaningful results for these years. In most cases we pay more attention to the periods 6–11. The US data do not need applying a cleaning procedure because trades are much more intensive.

We examine the changes of the stock prices during a week. In order to calculate the weekly return on the stock we take the close prices of the last trade days of the neighboring weeks. For example, if the trade takes place on each business day of the weeks we take the close prices of the examined stock on Fridays of the considered

Table 1 Characteristics of the considered time periods

Period #	Starting date	Ending date	Number of stocks (Russia)	Number of stocks after applying cleaning procedure (Russia)	Number of stocks (USA)
1	01/01/2001	12/30/2001	98	45	3332
2	12/31/2001	12/29/2002	64	50	3492
3	12/30/2002	12/28/2003	102	71	3637
4	12/29/2003	12/26/2004	95	63	3883
5	12/27/2004	12/25/2005	222	106	4150
6	12/26/2005	12/24/2006	448	241	4404
7	12/25/2006	12/23/2007	539	313	4739
8	12/24/2007	12/21/2008	650	276	4866
9	12/22/2008	12/20/2009	451	290	5029
10	12/21/2009	12/19/2010	463	305	5331
11	12/20/2010	12/18/2011	570	307	5607

and previous weeks. In order to calculate a weekly inflation rate we have made an assumption that the inflation rate is the same each day of the month. So we take the monthly inflation rate provided by the official statistic authorities divide it by the number of days in the month and so obtain the inflation rate for each day. Weekly inflation rate is evaluated as the sum of daily inflation rates of the examined week.

So we have determined the profitability for each stock in the following way. The stock is said to be profitable if it is traded during the week and its weekly revenue is higher than the inflation rate. If it is not traded during this particular week, then this stock is considered as not profitable without any calculations.

4 The Comparative Analysis of the Russian and US Stock Markets Models

4.1 Distribution of Performance Measure Values

The first subject of our analysis is the distribution of performance measure values between all pairs of stocks. Figures 1, 2, 3 show the distributions for Russian stock market; Figs. 4, 5, and 6 show distributions for US stock market. Table 2 shows the mean and standard deviation values.

A very unstable form can characterize the histograms for Russia in the time period 2001–2005. This can be explained by a small number of the stocks that are actively traded (less than 100). From the year 2005 the curves of Russian and US markets look more similar to the normal distribution. The edges with the weight more than 30 are rare for both of the countries. The curve of the USA stock market has a stable form in all time periods. The most weights of the edges in all time intervals vary from 10 till 20. In the time periods from 2006 to 2011 the curve for the Russian stock market changes much from period to period. The most significant

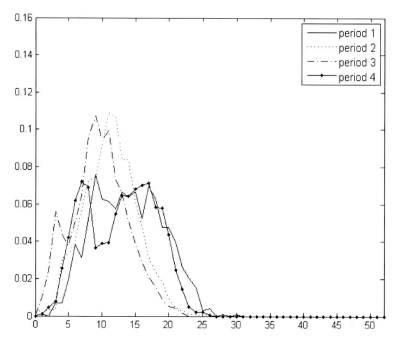

Fig. 1 Distributions of performance measure values of the Russian stock market for periods 1–4

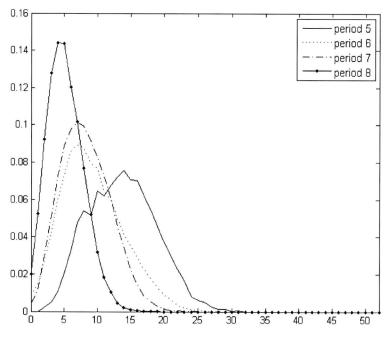

Fig. 2 Distributions of performance measure values of the Russian stock market for periods 5–8

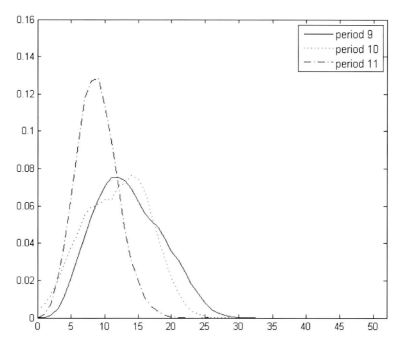

Fig. 3 Distributions of performance measure values of the Russian stock market for periods 9–11

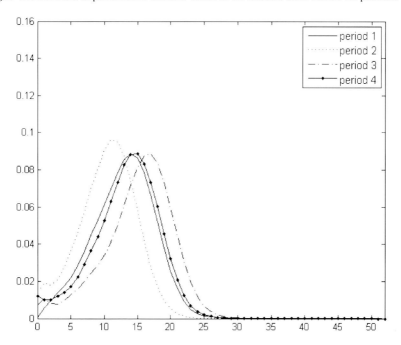

Fig. 4 Distributions of performance measure values of the US stock market for periods 1–4

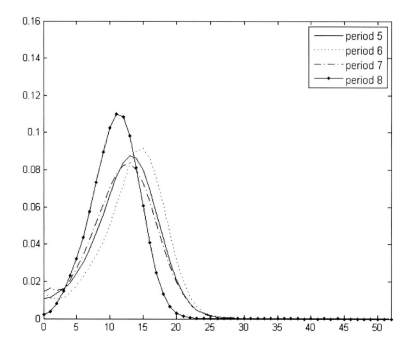

Fig. 5 Distributions of performance measure values of the US stock market for periods 5–8

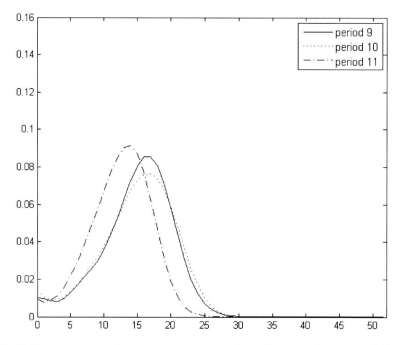

Fig. 6 Distributions of performance measure values of the US stock market for periods 9–11

Table 2 Mean and standard deviations values for considered periods

Period #	Mean value (Russia)	Standard deviation (Russia)	Mean value (USA)	Standard deviation (USA)	Change in MICEX Index	Change in S&P500 Index
1	13.9	5.1	12.9	4.6	64.58%	−9.53%
2	11.3	3.9	10.4	4.3	34.16%	−23.75%
3	9.7	4.2	15.0	5.2	62.35%	24.62%
4	12.8	5.1	13.3	4.9	5.71%	9.07%
5	13.7	5.0	12.3	4.8	85.69%	5.29%
6	9.2	4.6	13.2	4.9	59.40%	11.20%
7	8.2	3.8	11.9	5.0	19.25%	5.22%
8	5.2	2.7	10.8	3.7	−68.31%	−40.67%
9	13.5	5.1	14.9	5.2	123.58%	26.48%
10	11.8	4.9	14.9	5.6	21.81%	11.66%
11	8.8	3.1	12.4	4.5	−17.00%	−2.20%

changes for Russia happen in the crisis year 2008 and in 2011. The mean value of the performance measure for Russia also varies more than for the USA, which is quite stable. The standard deviation is stable for both countries stock markets (see Table 2).

It is interesting to compare these characteristics with the index changes in the same time periods. Performance measure and index show the profitability of the stock market so they should be related. In Table 2 there are mean values, standard deviations and also the values of the indexes MICEX and S&P500 measured in the same time intervals. We can see that both indexes fall in the years 2008 and 2011 and there is a huge growth of the MICEX index in the years 2005 and 2009. We can also notice that indexes show similar tendency during 2003–2011. In the years 2001 and 2002 index S&P500 shows a negative yield, while MICEX has a positive yield, however they both have the negative dynamics in this time period; the values for MICEX and S&P500 in the year 2002 are less than in the year 2001. We can compare the change of the mean value with the change of the index. For the US stock market there is a following rule: if the index goes down, the mean value also goes down and vice versa. The exclusions were periods 9 and 10, as during these time intervals the decrease of the index was significant, while the decrease of the mean value was relatively small (0.002). For the Russian stock market this pattern takes place from 2006. During the years 2001–2005 the change of the index was not accompanied by a corresponding change of the mean value.

All in all, the analysis of distributions and the tables of the mean values as well as the comparison of the performance measure value with the index change reveal a significant unsteadiness of the Russian stock market compared to the more developed market of the USA. It also shows a huge influence of the crisis year 2008 on the Russian market.

Table 3 Density of the Russian market graph for different threshold values

Threshold/Period #	1	2	3	4	5	6	7	8	9	10	11
10	0.77	0.68	0.51	0.68	0.78	0.43	0.35	0.07	0.76	0.67	0.40
11	0.71	0.59	0.42	0.64	0.71	0.36	0.27	0.04	0.69	0.60	0.28
12	0.64	0.48	0.32	0.60	0.65	0.29	0.20	0.02	0.62	0.54	0.19
13	0.59	0.37	0.24	0.55	0.58	0.23	0.14	0.01	0.54	0.47	0.12
14	0.52	0.29	0.18	0.48	0.51	0.18	0.09	0.00	0.47	0.40	0.07
15	0.46	0.20	0.13	0.42	0.44	0.14	0.06	0.00	0.40	0.32	0.04
16	0.39	0.14	0.09	0.35	0.36	0.11	0.03	0.00	0.34	0.25	0.02
17	0.34	0.09	0.06	0.28	0.29	0.08	0.02	0.00	0.28	0.18	0.01
18	0.27	0.06	0.04	0.21	0.23	0.05	0.01	0.00	0.23	0.12	0.01
19	0.21	0.04	0.02	0.15	0.18	0.04	0.01	0.00	0.18	0.08	0.00
20	0.16	0.02	0.01	0.09	0.13	0.02	0.00	0.00	0.14	0.05	0.00

Table 4 Density of the US market graph for different threshold values

Threshold/Period #	1	2	3	4	5	6	7	8	9	10	11
10	0.77	0.61	0.86	0.79	0.74	0.79	0.70	0.65	0.85	0.84	0.76
11	0.71	0.52	0.82	0.74	0.67	0.74	0.63	0.55	0.81	0.80	0.69
12	0.64	0.42	0.78	0.68	0.59	0.68	0.55	0.44	0.77	0.75	0.61
13	0.56	0.33	0.73	0.60	0.51	0.61	0.47	0.33	0.72	0.70	0.53
14	0.47	0.24	0.67	0.52	0.42	0.53	0.39	0.23	0.66	0.64	0.44
15	0.38	0.17	0.60	0.43	0.33	0.44	0.31	0.15	0.58	0.57	0.35
16	0.30	0.11	0.51	0.34	0.25	0.35	0.23	0.09	0.50	0.50	0.26
17	0.22	0.07	0.43	0.26	0.18	0.26	0.17	0.05	0.42	0.42	0.18
18	0.15	0.04	0.34	0.19	0.13	0.18	0.12	0.03	0.33	0.35	0.12
19	0.10	0.02	0.25	0.13	0.09	0.12	0.08	0.01	0.25	0.27	0.07
20	0.06	0.01	0.18	0.08	0.05	0.08	0.05	0.01	0.18	0.21	0.04

4.2 Edge Density and Degree Distribution

By fixing the threshold value we define the particular market graph. Edge density of an undirected graph is defined by (2)

$$D = \frac{2|E|}{|V|(|V|-1)} \quad (2)$$

where V is the set of vertices and E is the set of edges.

Tables 3 and 4 show the edge density of the Russian and US market graphs, respectively, for threshold values from 10 to 20. If we choose the threshold bigger than 20 or less than 10, then we will get too dense or too sparse market graph. Analysis of such graphs does not lead to the interesting results.

We can observe again that the Russian stock market has much more serious fluctuation than US stock market. It is difficult to compare the densities of Russian and US stock market graphs for the periods 1–5. It was mentioned before that the small number of stocks for these periods do not allow us to have reliable results. For the periods 6–11 for the majority of the threshold values the density of the corresponding market graph is higher for the USA than for Russia.

The most common observed phenomenon of the real life graphs is the following the power–law model. According to this model, the probability that the vertex has k edges emanating from it is given by (3)

$$P(k) \propto k^{-\gamma} \qquad (3)$$

or, equivalently

$$\log P(k) \propto -\gamma \log k \qquad (4)$$

Boginski et al. [4] show that the market graph constructed using the Pearson correlations similarity measure for US stock market for time period from 1998 to 2002 have well-defined power–law structure. Our experiments show that for some threshold values for all considered time periods the market graph with performance measure of the similarity has the tendency to be similar to power–law distribution. Figures 7 and 8 show the degree distributions (in the logarithmic scale) for some

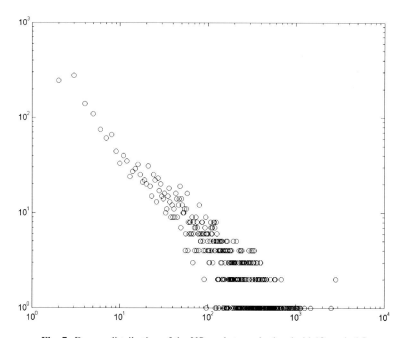

Fig. 7 Degree distribution of the US market graph, threshold 19, period 2

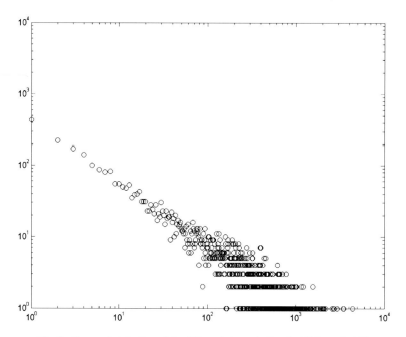

Fig. 8 Degree distribution of the US market graph, threshold 21, period 11

instances of the US market graph corresponding to different periods. These plots can be approximated by straight lines, which means that the distribution has the tendency to be similar to power–law distribution. If we examine all the periods and all the thresholds, then the graph shows the same behavior for all of the periods. For huge and small thresholds the graph is too sparse or too dense to judge what structure it has. For the graphs with the density less than 0.23 a straight line can be recognized for all time periods. This density can be obtained by setting the threshold in the range from 15 till 20. The most accurate straight line is obtained for the thresholds near 20. If we continue reducing the density of the edges the straight line can still be recognized until the moment, when there are not enough vertexes with different degrees. It is hard to check if the hypothesis about the power–law nature of the model is correct, because there are few stocks on the Russian stock market.

4.3 Clustering Coefficient

It is interesting to compare the clustering coefficients of the considered graphs. The local clustering coefficient C_i for a particular vertex i is given by the proportion of edges between the neighbors divided by the maximum possible number of edges

Table 5 Clustering coefficients for different threshold values for Russian stock market graphs

Threshold/Period #	1	2	3	4	5	6	7	8	9	10	11
10	0.90	0.85	0.78	0.86	0.92	0.81	0.76	0.42	0.89	0.88	0.76
11	0.89	0.77	0.74	0.81	0.89	0.74	0.71	0.34	0.88	0.86	0.72
12	0.88	0.75	0.67	0.79	0.86	0.69	0.65	0.22	0.87	0.81	0.69
13	0.83	0.73	0.62	0.78	0.83	0.63	0.58	0.15	0.83	0.77	0.64
14	0.75	0.72	0.60	0.72	0.77	0.60	0.52	0.09	0.81	0.73	0.56
15	0.72	0.61	0.47	0.69	0.73	0.54	0.42	0.03	0.77	0.70	0.43
16	0.66	0.52	0.40	0.64	0.71	0.49	0.37	0.02	0.74	0.66	0.30
17	0.63	0.39	0.35	0.61	0.69	0.39	0.29	0.01	0.71	0.59	0.22
18	0.60	0.29	0.27	0.56	0.66	0.32	0.23	0.00	0.64	0.53	0.14
19	0.57	0.16	0.16	0.46	0.60	0.27	0.14	0.00	0.56	0.48	0.06
20	0.49	0.12	0.13	0.33	0.50	0.24	0.08	0.00	0.52	0.42	0.03

between them. More precisely, if N_i is a neighborhood of the vertex v_i and E is the set of edges, then the local clustering coefficient is defined by

$$C_i = \frac{2|\{e_{jk} : v_j, v_k \in N_i, e_{jk} \in E\}|}{n} \qquad (5)$$

where k_i is the number of neighbors of a vertex v_i.

The clustering coefficient of the graph C is given as an average of the local clustering coefficients of all vertices.

$$C = \frac{1}{n}\sum_{i=1}^{n} C_i \qquad (6)$$

Tables 5 and 6 show the clustering coefficient values for different threshold values for Russian and US stock market graphs, respectively.

The analysis of the Tables 3, 4, 5, 6 shows that the clustering coefficient values for both considered market graphs are bigger than the edge density. The US market graph has the stable clustering coefficient for different time periods. These characteristics are typical for power–law graphs. The value of the clustering coefficient for Russian stock market changes much more significantly from period to period. The minimum value is observed during the crisis of the 2008. We would like to note that for US market graph the clustering coefficient also decreases during 2008 but this change is small. If we choose such thresholds for Russian and US stock market that the edge densities are the same, then the clustering coefficients values for US market graph are bigger than for Russian market graph.

We can also denote a tendency to the decrease of the index, density of the market graph and of the clustering coefficients in years 2005–2008 for Russian stock

Table 6 Clustering coefficients for different threshold values for US stock market graphs

Threshold/ Period #	1	2	3	4	5	6	7	8	9	10	11
10	0.91	0.84	0.94	0.91	0.89	0.92	0.88	0.86	0.94	0.93	0.90
11	0.89	0.81	0.93	0.90	0.86	0.90	0.85	0.82	0.93	0.92	0.88
12	0.87	0.78	0.91	0.87	0.84	0.87	0.83	0.79	0.91	0.90	0.86
13	0.84	0.75	0.89	0.84	0.81	0.84	0.81	0.75	0.89	0.89	0.83
14	0.81	0.72	0.87	0.81	0.79	0.81	0.79	0.72	0.87	0.86	0.81
15	0.78	0.70	0.84	0.79	0.77	0.79	0.78	0.70	0.84	0.84	0.78
16	0.75	0.69	0.81	0.76	0.76	0.76	0.77	0.69	0.82	0.82	0.75
17	0.72	0.68	0.78	0.75	0.75	0.74	0.76	0.67	0.80	0.80	0.73
18	0.70	0.67	0.76	0.73	0.74	0.72	0.75	0.65	0.78	0.78	0.70
19	0.69	0.67	0.74	0.72	0.73	0.71	0.73	0.61	0.77	0.76	0.68
20	0.68	0.65	0.73	0.71	0.71	0.70	0.71	0.55	0.76	0.74	0.64

market. If we choose for these periods such thresholds, that graphs have the same density, then the values of the clustering coefficients remain relatively stable (except the year 2008—for which it is hard to make any comparison).

5 Conclusions

The modification of the stock market graph model suggested in this study allows to evaluate the potential profitability, performance of the examined stock market. The analysis of the stock market graphs, built using the suggested similarity measure, showed that US stock market graphs for a number of thresholds are similar to the power–law model. For the Russian stock market it is not possible to check the hypothesis of the power–law nature of the model, as there are not enough stocks in the market.

According to the comparison of Russian and US stock markets we can denote that the values of characteristics are stable for US markets and that there are big changes of these values for Russia in certain time periods. For example, in the crisis year 2008 graph characteristics of Russian stock markets changed significantly, whereas US characteristics changed but not that much. In other periods characteristics of Russian stock market graph are close to the US market graph, however the changes in the index values and the amount of the trading stocks are different for the countries.

References

1. Bautin, G.A., Kalyagin, V.A., Koldanov, A.P., Koldanov, P.A., Pardalos P.M.: Simple measure of similarity for the market graph construction. Comput. Manag. Sci. **10**(2–3). 105–124 (2013)
2. Boginski, V., Butenko, S., Pardalos, P.M.: On structural properties of the market graph. In: Nagurney, A. (ed.) Innovations in Financial and Economic Networks, pp. 29–45. Edward Elgar Publishing, London (2003)
3. Boginski, V., Butenko, S., Pardalos, P.M.: Statistical analysis of financial networks. Comput. Stat. Data Anal. 431–443 (2005)
4. Boginski, V., Butenko, S., Pardalos, P.M.: Mining market data: A network approach. Comput. Oper. Res. 3171–3184 (2006)
5. Huang, W-Q, Zhuang, X-T, Shuang, Y.: A network analysis of the Chinese stock market. Phys. A **388**, 2956–2964 (2009)
6. Jallo, D., Budai, D., Boginski, V., Goldengorin, B., Pardalos, P.M.: Network-based representation of stock market dynamics: an application to american and swedish stock markets. In: Goldengorin, B., Kalyagin, V., Pardalos, P. (eds.) Models, Algorithms, and Technologies for Network Analysis, Springer Proceedings in Mathematics & Statistics, vol. 32, pp. 91–98. Springer, New York (2012)
7. Vizgunov, A., Glotov, A., Pardalos, P.M.: Comparative analysis of the BRIC countries stock markets using network approach. In: Goldengorin, B., Kalyagin, V., Pardalos, P. (eds.) Models, Algorithms, and Technologies for Network Analysis, Springer Proceedings in Mathematics & Statistics, vol. 59, pp. 191–201. Springer, New York (2013)
8. Vizgunov, A., Goldengorin, B., Kalyagin, V., Koldanov, A., Koldanov, P., Pardalos P.M.: Network approach for the Russian stock market. Comput. Manag. Sci. **11**(1–2), 45–55 (2014)

The Flatness Theorem for Some Class of Polytopes and Searching an Integer Point

Dmitry V. Gribanov

Abstract Let A be an $m \times n$ integral matrix of the rank n, we say that A has bounded minors if the maximum of the absolute values of the $n \times n$ minors is at most k, we will call these matrices as k-*modular*. We investigate an integer program $max\{c^\top x : Ax \leq b, x \in \mathbb{Z}^n\}$, where A is k-modular. We say that A is *almost unimodular* if it is 2-modular and the absolute values of its $(n-1) \times (n-1)$ minors are at most 1. We also refer 2-modular matrices to as *bimodular*. We say that A is *strict k-modular* if the absolute values of its $n \times n$ minors are from set $\{0, k, -k\}$. We prove that the width of an empty lattice polytope is less than $(k-1)(n+1)$ if it is induced by a system of inequalities with a *strict k-modular* matrix. Furthermore, we can give a polynomial-time algorithm for searching an integer point in a *strict k-modular* polytope if its width is grater than $(k-1)(n+1)$.

1 Introduction

Let A be an $m \times n$ integral matrix. Its ij-th element is denoted by A_{ij}, A_{i*} is the i-th row of A and A_{*j} is the j-th column of A. The set of rows (resp. columns) of A with numbers in I is denoted by A_{I*} (resp. by A_{*I}), $\Delta_k(A)$ is the maximum of the absolute values of the $k \times k$ minors of A. The rank of A is denoted by $rank(A)$. We say that A is k-modular if $\Delta_{rank(A)}(A) \leq k$, where $k \in \mathbb{N}$. Following the terminology of [17], we refer 2-modular matrices to as *bimodular*. The matrix A is called *almost unimodular* if $\Delta_{rank(A)}(A) = 2$ and $\Delta_{rank(A)-1}(A) \leq 1$ (see [4]). For a vector $b \in \mathbb{Z}^n$, by $P(A, b)$ (or by P) we denote the polyhedron $\{x \in \mathbb{R}^n : Ax \leq b\}$. Let u be a vertex of P, $I(u) = \{i : A_{i*}u = b_i\}$, $N(u) = \{x : A_{I(u)*}x \leq b_{I(u)}\}$. The set $N(u)$ is called the *corner polyhedron* associated with the vertex u.

For a matrix $B \in \mathbb{R}^{s \times n}$, $cone(B) = \{x : x = Bt, t \in \mathbb{R}^n, t_i \geq 0\}$ is the cone spanned by columns of B and $conv(B) = \{x : x = Bt, t \in \mathbb{R}^n, t_i \geq 0, \sum_{i=1}^{n} t_i = 1\}$ is the convex hull spanned by columns of B.

D.V. Gribanov (✉)
Nizhny Novgorod Lobachevsky State University, National Research University Higher School of Economics, Nizhny Novgorod, Russian Federation
e-mail: dimitry.gribanov@gmail.com

The following three theorems were proved by S.I. Veselov and A.J. Chirkov in the paper [17] (assuming $A \in \mathbb{Z}^{m \times n}$ is bimodular, $c \in \mathbb{Z}^n$, $b \in \mathbb{Z}^m$):

Theorem 1. *If $P(A, b)$ is full-dimensional, then $P(A, b) \cap \mathbb{Z}^n \neq \emptyset$.*

Theorem 2. *Let u be a vertex of $P(A, b)$. Then, each vertex of $conv(N(u) \cap \mathbb{Z}^n)$ lies on an edge of $P(A, b)$.*

Theorem 3. *If each $n \times n$ minor of A is not equal to zero, then the problem $\max\{c^\top x : Ax \leq b, \ x \in \mathbb{Z}^n\}$ can be solved in polynomial time.*

The work [17] contains the following simple corollaries:

Corollary 1. *One can check the emptiness of the set $P(A, b) \cap \mathbb{Z}^n$ in polynomial time.*

Corollary 2. *Let the augmented matrix $\begin{pmatrix} c^\top \\ A \end{pmatrix}$ be bimodular, then the problem $\max\{c^\top x : Ax \leq b, \ x \in \mathbb{Z}^n\}$ can be solved in polynomial time.*

The interesting result was obtained in [6] for *almost unimodular* matrices.

Theorem 4. *Let $A \in \mathbb{Z}^{m \times n}$ be almost unimodular, $b \in \mathbb{Z}^m$, $c \in \mathbb{Z}^n$. Then, there is a polynomial-time algorithm to solve $\max\{c^\top x : Ax \leq b, \ x \in \mathbb{Z}^n\}$.*

Some remarkable result was obtained by V.E. Alekseev and D.V. Zakharova in [1] for $\{0, 1\}$-matrices.

Theorem 5. *Let $A \in \{0, 1\}^{m \times n}$, $b \in \{0, 1\}^m$, $c \in \{0, 1\}^n$, for some fixed k the augmented matrix $\begin{pmatrix} c^\top \\ A \end{pmatrix}$ is k-modular and all rows of A have at most 2 units. Then, the problem $\max\{c^\top x : Ax \leq b, \ x \in \mathbb{Z}^n\}$ can be solved in polynomial time.*

Let $c \in \mathbb{Z}^n$, $M \subseteq \mathbb{R}^n$ and P be a convex body in $\mathbb{R}^n$. The *width of P with respect to c* (denoted by $width_c(P)$) is the difference $\max\{c^\top x : x \in P\} - \min\{c^\top x : x \in P\}$, $width_M(P) = \min_{c \in M}\{width_c(P)\}$. The *width of P* (denoted by $width(P)$) is $width_{\mathbb{Z}^n \setminus \{0\}}(P)$. Also, we need the set $Flat(P) = \{c \in \mathbb{Z}^n \setminus \{0\} : width_c(P) = width(P)\}$.

Now we refer to the classical flatness theorem due to Khinchine [12]. Let P be a convex body, Khinchine shows that if $P \cap \mathbb{Z}^n = \emptyset$, then $width(P) \leq f(n)$, where $f(n)$ is a value that depends only on the dimension. There are many estimates on the value of $f(n)$ in works [2, 3, 5, 9, 12, 13]. The conjecture claims that $f(n) = O(n)$. In this work we partially prove this conjecture for polytopes that are induced by systems with strict k-modular matrices, for $k = O(1)$ (in other words we fix a class of matrices with the property to be strict k-modular). The best known upper bound on $f(n)$ is $O(n^{4/3} \log^c(n))$ due to Rudelson [13], where c is some constant that does not depend on n.

An interesting problem is estimating the value $f(n)$ for empty lattice simplices [2, 8, 10, 15]. A simplex S is called empty lattice if $Vert(S) \subseteq \mathbb{Z}^n$ and $S \cap \mathbb{Z}^n \setminus Vert(S) = \emptyset$, where $Vert(S)$ is the set of vertices of S. The best known estimate of $f(n)$ for empty lattice simplices is $O(n \log(n))$ due to [2].

The work [6] contains an estimate of the width of an empty lattice simplex that is induced by a k-modular matrix.

Theorem 6. *Let $A \in \mathbb{Z}^{m \times n}$, $b \in \mathbb{Z}^m$, $P(A,b)$ be a simplex or a parallelotope, Δ_{min} and Δ_{max} be the absolute values of a minimal and maximal non-zero minors of A. If $P(A,b) \cap \mathbb{Z}^n \neq \emptyset$, then width$(P(A,b)) \leq (\Delta_{min} - 1)\Delta_{min}(n+1)$.*

Furthermore, one can give a polynomial-time algorithm for searching an integer point in a simplex if its width is greater than $(\Delta_{min} - 1)\Delta_{min}(n+1)$.

Our goal is proving analogs of the classical flatness theorem by considering the dimension and the absolute values of minors as the main parameters.

2 Results

Theorem 7. *Let $A \in \mathbb{Z}^{m \times n}$ is strict Δ-modular, $b \in \mathbb{Z}^m$, $P(A,b)$ is a polytope. If width$(P(A,b)) > (\Delta - 1)(n+1)$, then $|P(A,b) \cap \mathbb{Z}^n| \geq n+1$. An integer point in $P(A,b) \cap \mathbb{Z}^n$ can be found in polynomial time.*

We need the following propositions and lemmas to prove the Theorem 7.

Proposition 1. *If M and P are polytopes, $P \subseteq M$, then width$(P) \leq$ width(M).*

Proof. Obvious.

Let $C = cone(B)$ and B is a matrix. We define two sets: $\mathscr{F}(C) = conv((0\ B))$ and $\mathscr{G}(C) = conv(B) + cone(B)$. It is easy to see $C = \mathscr{F}(C) \cup \mathscr{G}(C)$.

Proposition 2. *Let $B \in \mathbb{Z}^{n \times s}$. If $v \in \mathscr{F}(C)$, then $\exists t \in \mathbb{R}^s : v = Bt, 0 \leq t_i \leq 1, \sum_{i=1}^s t_i \leq 1$.*

Proof. Obvious.

Lemma 8. *Let $B \in \mathbb{Z}^{n \times s}$, $c \in \mathbb{Z}^n$ and $x^* \in \mathbb{Z}^n$ be an optimal solution of the integer program $\min\{c^\top x : x \in C \cap \mathbb{Z}^n \setminus \{0\}\}$, then $x^* \in \mathscr{F}(C)$.*

Proof. All vertices of $\mathscr{G}(C)$ are integral. Hence, some vertex of $\mathscr{G}(C)$ is an optimal solution of $\min\{c^\top x : x \in \mathscr{G}(C) \cap \mathbb{Z}^n\}$. But, the vertices of $\mathscr{F}(C)$ and $\mathscr{G}(C)$ are coinciding, except $0 \in \mathscr{F}(C)$. So, $\min\{c^\top x : x \in C \cap \mathbb{Z}^n \setminus \{0\}\} = \min\{c^\top x : x \in \mathscr{F}(C) \cap \mathbb{Z}^n \setminus \{0\}\}$ and $x^* \in \mathscr{F}(C)$.

Let P be a polytope and v be a vertex of P. The *normal cone* of the vertex v (denoted by $H_P(v)$) is $\{c^\top \in \mathbb{R}^n : \{x : c^\top x \leq c^\top v\} \cap int(P) = \emptyset\}$, where $int(P)$ is the set of interior points of P. If $P = P(A,b)$ and $I(v) = \{i : A_{i*}v = b_i\}$, then $H_P(v) = cone(A_{I(v)*}^\top)$.

Lemma 9. *Let $P = P(A,b)$ be a polytope, width$(P) > 0$ and $c^* \in Flat(P)$. Then, there are vertices v, u of P, such that $c^* \in \mathscr{F}(H_P(v) \cap (-H_P(u)))$ and width$(P) = c^{*\top}(v - u)$.*

Proof. Let V be the set of vertices of P. Since $width(P) > 0$, then $\bigcup_{v \in V} H_P(v) = \mathbb{R}^n$. We can rewrite this formula as follows: $\mathbb{R}^n = \bigcup_{v \in V} \bigcup_{u \in V \setminus \{v\}} (H_P(v) \cap -H_P(u))$. Let us define $C_P(v, u) = H_P(v) \cap -H_P(u) \cap \mathbb{Z}^n \setminus \{0\}$. Then, $width(P) = min_{v \in V} min_{u \in V \setminus v} \{width_{C_P(u,v)}(P)\}$.

We have $width_{C_P(v,u)}(P) = min_{c \in C_P(v,u)} \{max\{c^T x : x \in P\} - min\{c^T x : x \in P\}\} = min_{c \in C_P(v,u)}\{c^T(v-u)\}$. Thus, $\exists u, v \in P$, $u \neq v$: $c^* \in C_P(u,v)$ and $width(P) = c^{*T}(v-u)$. Hence, by the Lemma 8: $c^* \in \mathscr{F}(H_P(v) \cap -H_P(u))$ for some vertices u, v.

Corollary 3. *Let A be strict Δ-modular, $P = P(A, b)$ be a polytope, $width(P) > 0$ and $c^* \in Flat(P)$. Then, there are matrices B_v, $B_u \in \mathbb{Z}^{n \times n}$ and vectors t_v, $t_u \in \mathbb{R}^n$, such that $c^* = B_v t_v$, $c^* = B_u t_u$ where $t_v \geq 0$, $t_u \geq 0$, $\sum_i t_{ui} + \sum_i t_{vi} \leq (n+1)$. The transposed columns of B_v and B_u are some linear independent rows of A.*

Proof. By the Lemma 9 we know that $c^* \in \mathscr{F}(H_P(v) \cap (-H_P(u)))$ for some vertices v, u of P. $H_P(v) = cone(A_{I(v)*}{}^T)$ and $H_P(u) = cone(A_{I(u)*}{}^T)$, where $I(v) = \{i : A_{i*}v = b\}$ and $I(u) = \{i : A_{i*}u = b\}$. Since $c^* \in H_P(v) \cap -H_P(u)$, by the Carathéodory's theorem [14, 18] we have that $c^* \in cone(B_v) \cap cone(B_u)$, where B_v and $-B_u$ are some $n \times n$ submatrices of $A_{I(v)*}{}^T$ and $A_{I(u)*}{}^T$. Moreover, $|det(B_v)| = |det(B_u)| = \Delta$.

Hence, c^* is a solution of the following program:

$$min\{c^T(v-u)\}$$
$$\begin{cases} c = B_v x = B_u y \\ c \in \mathbb{Z}^n \setminus \{0\} \\ x \geq 0, y \geq 0 \end{cases}$$

The program after rewriting is:

$$min\{x^T B_u^T(v-u)\} \qquad (1)$$
$$\begin{cases} B_v x = B_u y \\ B_v x \equiv 0 \,(mod\, 1) \\ x \geq 0, x \neq 0, y \geq 0 \end{cases}$$

Next we change the variables as follows $x \to x/\Delta$, $y \to y/\Delta$. It is easy to see that the program will be:

$$min\{x^T \frac{1}{\Delta} B_u^T (v-u)\} \qquad (2)$$

$$\begin{cases} B_v x = B_u y \\ B_v x \equiv 0 \,(mod\, \Delta) \\ x, y \in \mathbb{Z}_+^n,\, x \neq 0 \end{cases}$$

If we have an optimal solution of this program denoted by $\binom{x^*}{y^*}$, then $c^* = B_v x^* \frac{1}{\Delta} = B_u y^* \frac{1}{\Delta}$. Now we show that $\sum_i x^*_i + \sum_i y^*_i \leq \Delta(n+1)$. Let us consider the set $M = \{\binom{x}{y} : B_v x = B_u y, x \geq 0, y \geq 0\}$. This set is the cone in the $2n$ dimensional linear space. Let $d = dim(M)$, $d \leq n$. Any edge of M lies in the intersection of $d-1$ facets of M and $2n-d$ hyperplanes that induce a minimal linear subspace that includes M. Let g be an edge of M, then $g \in M \cap \{\binom{x}{y} : x_i = 0, y_j = 0, i \in I, j \in J\}$, where $I, J \subset \{1, 2, \ldots, n\}$ and $|I| + |J| = n - 1$. More precisely, $g \in \{z : Gz = 0\}$, where G is $n \times (n+1)$ submatrix of the matrix $(A_v - A_u)$ and $rank(G) = n$. Without loss of generality we can assume that the first n columns of G are linear independent. By this reason, we denote G_B as first n basis columns of G and G_N as the last column of G. Hence, $g \in \{\binom{z_B}{z_N} : G_B z_B = -G_N z_N\}$. Now, if we choose $z_N = 1$, then any component of z_B is a fraction of minors of matrix A, so it can be only 0 or 1. So, if $M \neq \emptyset$, then we can chose edge of M (denoted by g) such that g has at most $(n+1)$ unit components and all other components are 0-es. Next, we consider edge $g' = g\Delta$. It is easy to see that $g' \in M \cap \mathbb{Z}^{2n} \cap \{\binom{x}{y} : B_v x \equiv 0 \,(mod\, \Delta)\}$, so g' is a solution of the program (2). Let G' be the set of all edges that are created by the previous way, then by Lemma 8 we have that $\binom{x^*}{y^*} \in \mathscr{F}(cone(G'))$. By the Proposition 2 we conclude that $\sum_i x^*_i + \sum_i y^*_i \leq \Delta(n+1)$. Now, we return to the initial variables of program (1) and set $t_v = x^*/\Delta$, $t_u = y^*/\Delta$. Then $c^* = B_u t_u$ and $c^* = B_v t_v$.

Lemma 10. *Let the properties hold:*

1. $A \in \mathbb{Z}^{m \times n}$ is strict Δ-modular, $b \in \mathbb{Z}^m$;
2. $P = P(A, b)$ is a polytope;
3. $\delta \in \mathbb{R}^m$, $0 \leq \delta_i \leq 1$;

Then, $|width(P') - width(P)| = width(P) - width(P') \leq (n+1)$, *where* $P' = P(A, b - \delta)$.

Proof. By the Proposition 1 we have $|width(P') - width(P)| = width(P) - width(P')$.

First, we assume that $width(P') > 0$. Let $c^* \in Flath(P')$ and $c^* \in H_{P'}(q) \cap -H_{P'}(g)$ for some vertices q, g of P'. Then $width(P') = c^{*\top}(q - g)$. The $width(P) > 0$ and P is bounded, so $width(P) \leq c^{*\top}(v - u)$ for some vertices v, u of P, and $c^* \in H_P(v) \cap -H_P(u)$. By the Corollary 3 there are matrices $B_q, B_g \in \mathbb{Z}^{n \times n}$ and vectors $t_q, t_g \in \mathbb{R}^n$, such that $c^* = B_q t_q$, $c^* = B_g t_g$ where $t_q \geq 0$, $t_g \geq 0$, $\sum_i t_{q_i} + \sum_i t_{g_i} \leq (n+1)$. We know that $c^{*\top} q = t_q^\top B_q^\top q = t_q^\top (b - \delta)_{I_q}$, where I_q is the subset of the set $I(q) = \{i : A_{i*} q = b_i - \delta_i\}$, $|I_q| = n$. We make the same for the vertex g: $c^{*\top} g = t_g^\top B_g^\top g = -t_g^\top (b - \delta)_{I_g}$, where I_g

is the subset of the set $I(g) = \{i : A_i *g = b_i - \delta_i\}$, $|I_g| = n$. Hence, $width(P') = t_q^\top(b-\delta)_{I_q} + t_g^\top(b-\delta)_{I_g}$. Now we consider the vertices v, u. First we need to estimate value of $c^{*\top}v = t_q^\top B_q^\top v$. Rows of matrix $B_q^\top$ are rows of matrix A and lines of the system $B_q^\top x = B_q^\top v$ are not supporting hyperplanes of P. But the intersection of $B_q^\top x = B_q^\top v$ and P is not empty, so we can estimate $B_q^\top v$ as b_{I_q} and $t_q^\top B_q^\top v$ as $t_q^\top b_{I_q}$. Now we make the same for vertex u: $-c^{*\top}u = -t_g^\top B_g^\top u \leq t_g^\top b_{I_g}$. Hence, $width(P) - width(P') \leq c^{*\top}(v-u) - c^{*\top}(q-g) < (t_q^\top b_{I_q} + t_g^\top b_{I_g}) - (t_q^\top(b-\delta)_{I_q} + t_g^\top(b-\delta)_{I_g}) = t_q^\top \delta_{I_q} + t_g^\top \delta_{I_g} \leq (n+1)$.

Second, we consider the situation when $width(P') = 0$. The case of $width(P) = 0$ is trivial, so we assume that $width(P) > 0$. Then there exists $0 \leq \delta' < \delta$, $\delta' \neq 0$, such that for any $\epsilon > 0$ one has $width(P(A, b - \delta')) = 0$, but $width(P(A, b - \delta' + \bar{1}\epsilon)) > 0$. By the previous, if ϵ is small enough one has that $width(P) - width(P(A, b - \delta' + \bar{1}\epsilon)) \leq (n+1)$. Hence, $width(P) - width(P') = width(P) = width(P) - width(P(A, b - \delta' + \bar{1}\epsilon)) + width(P(A, b - \delta' + \bar{1}\epsilon)) \leq (n+1) + width(P(A, b - \delta' + \bar{1}\epsilon))$. But $width(P(A, b - \delta' + \bar{1}\epsilon))$ vanishes with ϵ goes to 0.

Now, we are ready to prove the main result (Theorem 7) of this section.

Proof. Any $x \in \mathbb{Z}$ can be represented as $x = y\lfloor x/y \rfloor + x \bmod y$, where $y \in \mathbb{Z}$ and $0 \leq x \bmod y < y$. Hence, $b_i = \Delta \lfloor b_i/\Delta \rfloor + b_i \bmod \Delta$. Let $b' \in \mathbb{Z}^m$, such that $b'_i = b_i - b_i \bmod \Delta$ and $P' = P(A, b')$. If $width(P) > (n+1)(\Delta - 1)$, then by the Lemma 10, $width(P') > 0$. Thus, P' is full-dimensional and each component of b' is divided by Δ. So, it is easy to see that P' is at least a simplex and all components of any vertex of P' are integer. Now the theorem follows from the fact that $P' \subset P$.

We can use any polynomial algorithm of a linear programming (Khachiyan's algorithm [11]) to find some vertex of P' as integer point of P.

Acknowledgements The author wishes to express special thanks for the invaluable assistance to A.J. Chirkov, S.I. Veselov, D.S. Malyshev, V.N. Shevchenko, and S.V. Sorochan. The work is partly supported by National Research University Higher School of Economics, Russian Federation Government grant, N. 11.G34.31.0057.

References

1. Alekseev, V.E., Zakharova, D.V.: Independent sets in graphs with bounded minors of the extended incidence matrix. Discrete Anal. Oper. Res. **17**(1), 3–10 (2010) [in russian]
2. Banaszczyk, W., Litvak, A.E., Pajor, A., Szarek, S.J.: The flatness theorem for non-symmetric convex bodies via the local theory of Banach spaces. Math. Oper. Res. **24**(3), 728–750 (1999)
3. Banaszczyk, W.: Inequalities for convex bodies and polar reciprocal lattices in $\mathbb{R}^n$ II: Application of K-convexity. Discrete Comput. Geom. **16**(3), 305–311 (1996)
4. Cornuéjols, G., Zuluaga, L.F.: On Padberg's conjecture about almost totally unimodular matrices. Oper. Res. Lett. **27**(3), 97–99 (2000)
5. Dadush, D.: Transference Theorems in the Geometry of Numbers. http://cs.nyu.edu/courses/spring13/CSCI-GA.3033-013/lectures/transference.pptx

6. Gribanov, D.V.: On Integer Programing With Almost Unimodular Matrices and The Flatness Theorem for Simplexes. Preprint (2014)
7. Grossman, J.W., Kilkarni, D.M., Schochetman, I.E.: On the minors of an incidence matrix and its Smith normal form. Linear Algebra Appl. **218**, 213–224 (1995)
8. Haase, C., Ziegler, G.: On the maximal width of empty lattice simplices. Eur. J. Combinatorics **21**, 111–119 (2000)
9. Kannan, R., Lovász, L.: Covering minima and lattice-point-free convex bodies. Ann. Math. **128**, 577–602 (1988)
10. Kantor, J.M.: On the width of lattice-free simplexes. Cornell University Library (1997) http://arxiv.org/abs/alg-geom/9709026v1
11. Khachiyan, L.G.: Polynomial algorithms in the linear programming. Comput. Math. Math. Phys. **20**(1), 53–72 (1980)
12. Khinchine, A.: A quantitative formulation of Kronecker's theory of approximation. Izvestiya Akademii Nauk SSR Seriya Matematika **12**, 113–122 (1948) [in russian]
13. Rudelson, M.: Distances between non-symmetric convex bodies and the MM^*-estimate. Positivity **4**(2), 161–178 (2000)
14. Schrijver, A.: Theory of Linear and Integer Programming. WileyInterscience Series in Discrete Mathematics. Wiley (1998)
15. Sebö, A.: An introduction to empty lattice simplexes. In: Cornuéjols, G., Burkard, R.R., Woeginger, R.E. (eds.) LNCS, vol. 1610, 400–414 (1999)
16. Shevchenko, V.N.: Qualitative Topics in Integer Linear Programming (Translations of Mathematical Monographs). AMS (1996)
17. Veselov, S.I., Chirkov, A.J.: Integer program with bimodular matrix. Discrete Optim. **6**(2), 220–222 (2009)
18. Ziegler, G.: Lectures on Polytopes, vol. 152. GTM, Springer,New York/Berlin/Heidelberg (1996)

How Independent Are Stocks in an Independent Set of a Market Graph

Petr A. Koldanov and Ivan Grechikhin

Abstract The problem of testing hypothesis of independence of random variables describing stock returns for a given set of stocks is considered. Two tests of independence are compared. The first test is the classical maximum likelihood test based on the determinant of a sample covariance matrix. The second test is the pairwise test used for market graph construction. This test is based on testing of pairwise independence of random variables describing stock returns by Pearson correlation test. The main result is the following: the maximum likelihood test is more powerful for a wide class of alternatives. Some examples are given.

1 Introduction

The problem of testing hypothesis of independence of random variables describing stock returns for a given set of stocks is considered in the paper. In particular this problem appears in analysis of a stock market. The main assumption we accept in this paper is that the joint distribution of random variables describing stock returns is the multivariate normal distribution. Consequently, to solve this problem it is possible to apply the classical maximum likelihood test based on the determinant of a sample covariance matrix [1]. However, this test is not usually used for the problems of the market graph analysis [2–6]. In common practice of stock market analysis a graph is used as a model of a stock market. The vertices of the graph represent the stocks, and the edges are added when the dependence between the corresponding stocks is sufficiently high.

The Pearson correlation is a popular measure of dependence between stocks [2, 9]. Based on this measure different structures such as a minimal spanning tree, a planar maximally filtered graph, a market graph, cliques and independent sets of a market graph are investigated [3, 4, 10, 11]. There exist different ways of a market graph construction. In particular, in the papers [5, 6] an edge between two vertices

---

P.A. Koldanov (✉) • I. Grechikhin
National Research University Higher School of Economics, Bolshaya Pecherskaya, 25/12, Nizhny Novgorod 603155, Russia
e-mail: pkoldanov@hse.ru

is added to a market graph if the corresponding pairwise Pearson correlation is not equal to zero. These authors call this graph a correlation network. In this case if the joint distribution of random variables describing stock returns is the multivariate normal distribution, then the definition of independence of random variables in a given set and the definition of an independent set in a graph (a set of pairwise non-adjacent vertices) are equivalent. This means that the hypothesis that the given stocks constitute an independent set in a market graph is equivalent to the hypothesis of independence of the corresponding random variables. Therefore this hypothesis can be tested by known statistical tools, such as tests of independence from the multivariate statistical analysis.

Another way of a market graph construction is suggested in [2–4]. An analysis of this procedure from the mathematical statistics point of view is performed in [7]. In the latter paper the concepts of a true market graph and a sample market graph are introduced. The construction procedure for a true (sample) market graph is the following: each vertex of a graph represents a stock; an edge between vertices i and j is added to a true (sample) market graph, if the true (sample) Pearson correlation between returns of stocks i and j is greater than a given threshold. The errors of identification of a true market graph by a sample market graph are investigated.

In the same way for both ways of a market graph construction one can introduce the concepts of a true independent set and a sample independent set. A true independent set is an independent set in a true market graph. A sample independent set is an independent set in a sample market graph. Then the procedure of a sample independent set construction can be considered as a statistical procedure of identifying an unknown true independent set. But this statistical procedure is not the only one that can be considered for identifying a true independent set. Moreover it is not clear whether this procedure is the best possible or even if this procedure is good from the statistical point of view.

In the present paper two tests of independence are compared. The first test is the classical maximum likelihood test based on the determinant of a sample covariance matrix [1]. The second test is the pairwise test used for market graph construction. This test is based on testing of pairwise independence of stocks by the Pearson correlation test. The main result is the following: the maximum likelihood test is more powerful for a wide class of alternatives.

The paper is organized as follows. In Sect. 2 the main definitions are introduced and the hypothesis of independence is formulated. In Sect. 3 two tests of independence are described. In Sect. 4 the power of the two tests for different alternatives is compared. In Sect. 6 the main results of paper are discussed.

2 Statement of the Problem

Let us consider a financial market with N stocks during n-day period of observations. We denote by $p_i(t)$ the price of the stock i for the day t ($i = 1, \ldots, N; t = 1, \ldots, n$) and define the daily return of the stock i for the period from $(t-1)$

to t as $r_i(t) = \ln(p_i(t)/p_i(t-1))$. We suppose $r_i(t)$ to be an observation of the random variable $R_i(t)$ and accept the following standard assumptions: the random variables $R_i(t), t = 1,\ldots,n$ are independent for fixed i, have the same distribution as a random variable R_i'' ($i = 1,\ldots,N$), and the random vector $(R_1, R_2, \ldots, R_N)$ has a multivariate normal distribution with covariance matrix $\|\sigma_{i,j}\|$. The sample covariance between the stocks i and j is defined by

$$s_{ij} = \frac{1}{n}\sum_{t=1}^{n}(r_i(t) - \overline{r_i})(r_j(t) - \overline{r_j}) \tag{1}$$

where $\overline{r_i} = \frac{1}{n}\sum_{t=1}^{n} r_i(t)$. It is known [1] that for a multivariate normal vector the statistics $(\overline{r_1}, \overline{r_2}, \ldots, \overline{r_N})$, $\|s_{ij}\|$ (where $\|s_{ij}\|$ is the sample covariance matrix) are sufficient. Therefore optimal statistical procedures can be based on these statistics only.

According to [7] matrix $\|\rho_{ij}\|$ with elements

$$\rho_{ij} = \frac{\sigma_{ij}}{\sqrt{\sigma_i \sigma_j}} \tag{2}$$

is a basic matrix for the construction of a true graph and matrix $\|r_{ij}\|$ with elements

$$r_{ij} = \frac{\sum(r_i(t) - \overline{r_i})(r_j(t) - \overline{r_j})}{\sqrt{\sum(r_i(t) - \overline{r_i})^2}\sqrt{\sum(r_j(t) - \overline{r_j})^2}} = \frac{s_{ij}}{\sqrt{s_{ii}s_{jj}}} \tag{3}$$

is a basic matrix for the construction of a sample graph.

Each vertex of the graph corresponds to a stock of the financial market. There are different procedures which can be used for construction of the graph. For example, in [2–4] the edge between two vertices i and j is included in a true market graph, if $\rho_{ij} > \rho_0$ (where ρ_0 is a threshold). In [5,6] an edge between two vertices i and j is added to a true graph, if $|\rho_{ij}| > 0$.

Below we consider the procedure of a market graph construction proposed in [5, 6]. In this case an independent set of vertices of a graph is a set of vertices such that every two vertices in the set are not connected by an edge, or $\rho_{ij} = 0$, $i, j \in \{i_1, \ldots, i_k\}$. This means that random variables R_l for $l \in \{i_1, \ldots, i_k\}$ are independent in the probabilistic sense.

In the present paper we consider the problem of testing the hypothesis of independence

$$H : \rho_{ij} = 0, i, j \in \{i_1, \ldots, i_k\} \tag{4}$$

for a given subset $R_{i_1}, \ldots, R_{i_k}$ of random variables $R_1, \ldots, R_N : (i_1, \ldots, i_k) \subset \{1, \ldots, N\}$.

3 Two Tests of Independence

Two tests of independence are considered in this section. The first test is the most popular test for a market graph construction. We call it simple test of independence. The second test is the classical maximum likelihood test from the multivariate analysis [1].

3.1 Simple Test of Independence

Let us test the hypothesis $h_{ij} : \rho_{ij} = 0$ for each pair of random variables $R_i, R_j : i, j \in \{i_1, \ldots, i_k\}$. If for all pairs of random variables from $R_{i_1}, \ldots, R_{i_k}$ the hypotheses $h_{ij} : \rho_{ij} = 0$ are accepted, then random variables $R_{i_1}, \ldots, R_{i_k}$ constitute an independent set.

The test of the hypothesis $h_{ij} : \rho_{ij} = 0$ is:

$$\varphi(x) = \begin{cases} 1, & r_{ij} > c_\alpha \\ 0, & r_{ij} \leq c_\alpha \end{cases} \tag{5}$$

where r_{ij} is given by (3) and constant c_α is a α-quantile of Student-distribution with $n - 2$ degree of freedom.

In this case condition "for all pairs of random variables from $R_{i_1}, \ldots, R_{i_k}$ the hypotheses $h_{ij} : \rho_{ij} = 0$ are accepted" can be rewritten as "$\max_{i,j \in \{i_1, \ldots, i_k\}} r_{ij} < c_\alpha$". Then the simple test of independence is:

$$\phi_{simple}(x) = \begin{cases} 1, & \max_{i_1, \ldots, i_k} r_{ij} > c_\alpha \\ 0, & \max_{i_1, \ldots, i_k} r_{ij} \leq c_\alpha \end{cases} \tag{6}$$

Test (6) can be considered as an application of the theory of multiple hypothesis testing [8].

3.2 Classical Test of Independence

It is known [1] that classical test of independence of random variables $R_{i_1}, \ldots, R_{i_k}$ is based on statistics

$$V = \frac{\det |s_{ij}|}{\prod_{i=1}^{K} s_{ii}} = \frac{\det |r_{ij}|}{\prod_{i=1}^{K} r_{ii}} \tag{7}$$

$i, j \in \{i_1, \ldots, i_k\}$.

When the hypothesis of independence of random variables $R_{i_1}, \ldots, R_{i_k}$ is true it is known that

$$P(-m \ln(V) \leq v) = P(\chi_f^2 \leq v) + O(m^{-2})$$

where for our case

$$f = \frac{k(k-1)}{2}; m = n - \frac{2k+11}{6}$$

and χ_f^2 is a random variable which has the χ^2-distribution with f degrees of freedom.

Then the classical test of independence is:

$$\phi_{classic}(x) = \begin{cases} 1, & -m \ln(V) < \chi_{f,\alpha}^2 \\ 0, & -m \ln(V) \geq \chi_{f,\alpha}^2 \end{cases} \quad (8)$$

where $\chi_{f,\alpha}^2$ is α-quantile of the χ^2-distribution with f degrees of freedom.

4 Comparison Analysis

In the present section the efficiency of the two tests of independence is compared. The efficiency of a test is defined as the minimal number of observations needed to reach power of at least 0.9. The power is defined as the probability of rejecting the hypothesis in the case when it is false. The four classes of alternatives are considered.

1. The first class: Correlation matrix has the following form:

$$M_{N,N} = \begin{bmatrix} 1 & \varepsilon & \varepsilon & \cdots & \varepsilon \\ \varepsilon & 1 & \varepsilon & \cdots & \varepsilon \\ \varepsilon & \varepsilon & 1 & \cdots & \varepsilon \\ \vdots & \vdots & \vdots & \ddots & \vdots \\ \varepsilon & \varepsilon & \varepsilon & \cdots & 1 \end{bmatrix}$$

The matrix of the class is constructed in such a way that each correlation coefficient in the matrix is equal to some value (epsilon). Since the epsilon is chosen close to zero (for example, 0.1, 0.2 or 0.3), this class of correlation matrices represents the situation, when every pair of random variables or stocks in the set has weak dependency.

2. The second class: Correlation matrix has the following form:

$$M_{N,N} = \begin{bmatrix} 1 & \varepsilon & 0 & \cdots & 0 \\ \varepsilon & 1 & 0 & \cdots & 0 \\ 0 & 0 & 1 & \cdots & 0 \\ \vdots & \vdots & \vdots & \ddots & \vdots \\ 0 & 0 & 0 & \cdots & 1 \end{bmatrix}$$

The matrix consists of correlation coefficients which are equal to zero, except one coefficient. The value for this distinct coefficient is chosen close to 1 (for example, 0.7, 0.8, 0.9). In other words, if we exclude one of the correlated stocks from our set of random variables, then the rest of the set will be independent.

3. For the third class the correlation matrix has the following form: all elements of the matrix are zero except seven diagonals: main diagonal with ones, than three pairs of diagonals where elements are equal to 0.3, 0.2, and 0.1. For example, a matrix which belongs to the third class is:

$$M_{7,7} = \begin{bmatrix} 1 & 0.3 & 0 & 0.2 & 0 & 0.1 & 0 \\ 0.3 & 1 & 0.3 & 0 & 0.2 & 0 & 0.1 \\ 0 & 0.3 & 1 & 0.3 & 0 & 0.2 & 0 \\ 0.2 & 0 & 0.3 & 1 & 0.3 & 0 & 0.2 \\ 0 & 0.2 & 0 & 0.3 & 1 & 0.3 & 0 \\ 0.1 & 0 & 0.2 & 0 & 0.3 & 1 & 0.3 \\ 0 & 0.1 & 0 & 0.2 & 0 & 0.3 & 1 \end{bmatrix}$$

Here we can name diagonals with coefficients as the first, the third, and the fifth or $N1=1$, $N2=3$, $N3=5$—diagonal's numbers with coefficients 0.3, 0.2, and 0.1, respectively. In our examples, N1 is always 1, therefore it is reasonable to name the amount of coefficients in this diagonal (which is $N-1$ for matrix of size N). This class of matrices represents the situation, when some of the stocks or random variables are weakly correlated (Figs. 1, 2, 3, and 4).

4. For the fourth class the correlation matrices are calculated from real market data, for example:

$$M_{5,5} = \begin{bmatrix} 1.0000 & 0.0634 & 0.0064 & -0.1642 & 0.0993 \\ 0.0634 & 1.0000 & 0.0337 & 0.5756 & 0.2437 \\ 0.0064 & 0.0337 & 1.0000 & 0.1341 & -0.0917 \\ -0.1642 & 0.5756 & 0.1341 & 1.0000 & 0.0428 \\ 0.0993 & 0.2437 & -0.0917 & 0.0428 & 1.0000 \end{bmatrix}$$

The stocks for this matrix are taken randomly. The comparison of tests is conducted for different real data correlation matrices.

How Independent Are Stocks in an Independent Set of a Market Graph 51

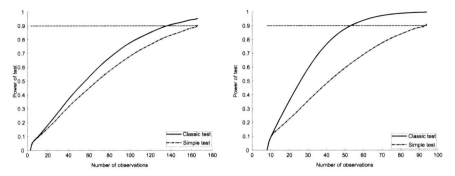

Fig. 1 Power functions of the two tests for a given number of observations. The covariance matrix from the first class with $\varepsilon = 0.2$, N=3 (*left*) and N=8 (*right*)

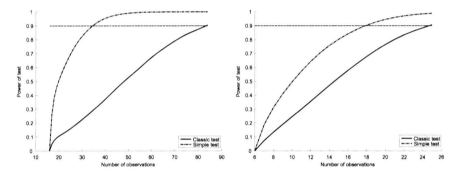

Fig. 2 Power functions of the two tests for a given number of observations. The covariance matrix from the second class with $\varepsilon = 0.7$, N=16 (*left*) and $\varepsilon = 0.8$, N=6 (*right*)

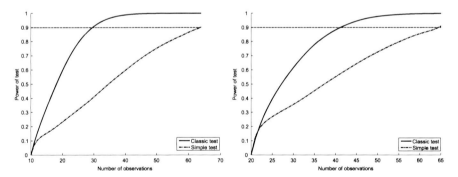

Fig. 3 Power functions of the two tests for a given number of observations. The covariance matrix from the third class with N=10, the number of coefficients for the first diagonal is 9, N2=4, N3=7 (*left*) and N=20, the number of coefficients for the first diagonal is 19, N2=2, N3=3 (*right*)

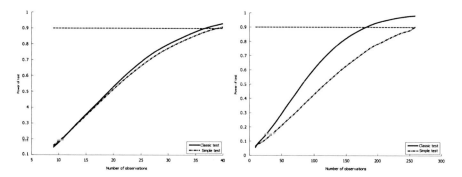

Fig. 4 Power functions of the two tests for a given number of observations. The covariance matrices from the fourth class are given in Sect. 4. The power of the classic test is higher for both cases

5 Numerical Results

In this section, the results of comparison of power for both tests are presented. A covariance matrix, matrices of the first, the second, the third, and the fourth classes are used. It may be noticed that the classic test shows better efficiency for the first, third, and fourth classes of correlation matrices, but the simple test is better for the second class of correlation matrices. It can be explained as follows: the second class matrices contain one non-zero coefficient with value that is close to one, and this coefficient influences the simple test more, because the simple test checks the coefficient value with a threshold. And vice versa, when we take the correlation matrix from the first class, small coefficients influence the determinant more and therefore the classic test is better for this class of matrices. It seems that the third and the fourth classes have a structure similar to the first class. The behavior of power functions for both tests for real data correlation matrices usually has two alternatives: both power functions have high values for small number of observations (more than 0.9), or the classic test power function has greater value.

6 Conclusion

The concept of an independent set in a graph and the concept of probabilistic independence of several random variables can be equivalent. This equivalence allows to apply classical methods of multivariate statistical analysis for an independent set construction. In the present paper it is shown that the maximum likelihood test is more powerful for a typical class of alternatives in particular for real market data. However, there exists a counterexample, in which the simple pairwise test is more powerful. Note that this counterexample has a theoretical interest only. Besides the obtained results can be applied for additional testing of independence of stocks included in an independent set of a market network.

Acknowledgements The authors are partly supported by National Research University Higher School of Economics, Russian Federation Government Grant N. 11.G34.31.0057 and RFFI Grant 14-01-00807.

References

1. Anderson, T.W.: An Introduction to Multivariate Statistical Analysis, 3rd edn. Wiley Interscience, New York (2003)
2. Boginsky, V., Butenko, S., Pardalos, P.M.: On structural properties of the market graph. In: Nagurney, A. (ed.) Innovations in Financial and Economic Networks, pp. 29–45. Edward Elgar Publishing, Northampton (2003)
3. Boginski, V., Butenko, S., Pardalos, P.M.: Statistical analysis of financial networks. J. Comput. Stat. Data Anal. **48**(2), 431–443 (2005)
4. Boginski, V., Butenko, S., Pardalos, P.M.: Mining market data: a network approach J. Comput. Oper. Res. **33**(11), 3171–3184 (2006)
5. Emmert-Streib, F., Dehmer, M.: Identifying critical financial networks of DJIA: towards a network based index. Complexity **16**, 1 24–33 (2010a)
6. Emmert-Streib, F., Dehmer, M.: Influence of the time scale on the construction of financial networks. PLoS ONE **5**, 9 (2010b)
7. Koldanov, A.P., Koldanov, P.A., Kalyagin, V.A., Pardalos, P.M.: Statistical procedures for the market graph construction. Comput. Stat. Data Anal. **68**, 17–29 (2013)
8. Lehmann, E.L., Romano, J.P.: Testing Statistical Hypotheses. Springer, New York (2005)
9. Mantegna, R.N.: Hierarchical structure in financial market. Eur. Phys. J. B **11**, 193–197 (1999)
10. Tumminello, M., Aste, T., Matteo, T.D., Mantegna, R.N.: A tool for filtering information in complex systems. Proc. Natl. Acad. Sci. **102**(30), 10421–10426 (2005)
11. Tumminello, M., Lillo, F., Mantegna, R.N.: Correlation, hierarchies and networks in financial markets. J. Econ. Behav. Organ. **75**, 40–58 (2010)

Analysis of Russian Industries in the Stock Market

Nina N. Lozgacheva and Alexander P. Koldanov

Abstract Analysis of industries in the stock market using graph theoretic model is performed. It is shown that in the Russian stock market there are some big industries which form sets that do not have any connection between each other and can be called clusters. The result is obtained not only by conventional approach to analyse the stock markets such as studying histograms, finding cliques and independent sets, but also by a more detailed investigation which is related, in particular, with the degrees of vertices. This analysis could provide division of the market to some sets of stocks which constitute industries in the stock market.

1 Introduction

One of the approaches to analyse a stock market is connected with a graph theoretic model which is called a market graph [1–3]. This analysis includes finding of different structures in a market graph such as cliques, independent sets, etc. In this paper we analyse industries in the Russian stock market using this approach.

The main question studied in this paper is to check whether the stocks of each of the considered industries form sets in the market graph that do not have any connection between each other and can be called clusters. To answer this question we order the stocks with respect to industry which leads to adjacency matrix of block structure. Next, we study histograms of the correlation coefficients, find cliques and independent sets and analyse cluster structure for each industry separately. To analyse cluster structure we consider the number of inter/intra-cluster edges with high weight (0.5; 0.6; 0.7). The main result is that the stocks of each of the considered industries in the Russian stock market form the corresponding clusters.

The paper is organized as follows. In Sect. 2 notions and definitions are introduced. In Sect. 3 the main industries in the Russian market are analysed.

N.N. Lozgacheva (✉) • A.P. Koldanov
LATNA Laboratory, National Research University Higher School of Economics,
Bolshaya Pecherskaya, 25/12, Nizhny Novgorod 603155, Russia
e-mail: nlozgacheva@hse.ru; akoldanov@hse.ru

In Sect. 4 the clustering structure of Russian market is discussed. In Sect. 5 concluding remarks are given.

2 General Concepts and Definitions

Let $G = (V, E)$ be an undirected graph with a set of vertices $V = \{1, \ldots, N\}$ and a set of edges $E = \{(i, j) : i, j \in V\}$. Let N be the number of stocks and n be the number of observations. Let $P_i(t)$ be the price of stock i, ($i = 1, \ldots, N$), on the day t, ($t = 1, \ldots, n$). Then $R_i(t) = \ln(\frac{P_i(t)}{P_i(t-1)})$ defines the log-return of the stock i over one-day period from $(t-1)$ until t. Correlation coefficients between returns of stocks i and j are calculated as:

$$r_{ij} = \frac{\sum_{i=1}^{n}(R_i(t) - \overline{R}_i)(R_j(t) - \overline{R}_j)}{\sqrt{\sum_{i=1}^{n}(R_i(t) - \overline{R}_i)^2}\sqrt{\sum_{i=1}^{n}(R_i(t) - \overline{R}_i)^2}}$$

where $\overline{R}_i = \frac{1}{n}\sum_{i=1}^{n} R_i(t)$ is the mean value of R_i. We use the matrix $\|r_{ij}\|$ to construct the market graph. For constructing the market graph we use the following procedure. Vertices represent stocks of the financial market and two vertices i and j are connected by an edge (i, j) if $r_{ij} \geq r_0$ where r_0 is threshold value, $-1 \leq r_0 \leq 1$. Different values of threshold r_0 define the market graphs with the same set of vertices and the different set of edges. The constructed graph is called a market graph.

A clique in a graph is a subgraph in which each pair of vertices is connected by an edge. A maximum clique is a clique that has the maximum possible size in the graph. A clique in a stock market is considered to be a strongly correlated group of stocks. An independent set is a subgraph in which every two vertices are not connected with an edge that can be easily interpreted for a stock market as a group of stocks that may constitute a diversified portfolio.

3 Analysis of Industries in the Russian Market

In this paper we used data of the Russian stock market during the period of four years (from September 1, 2009 to September 1, 2013). The prices of the stocks were taken from an open source—website of investment company Finam [4].

We analyse 105 companies in the Russian stock market. These companies belong to three industries: Mining and Quarrying (MQ); Electricity, Gas and Water Supply (EG); Chemicals and Chemical Products (CP) [5]. The lists of companies included in each industry are given in appendix. All these companies were traded during the analysed period.

In this section we study main characteristics (distribution of correlation coefficients, edge density, cliques and independent sets) of each industries separately.

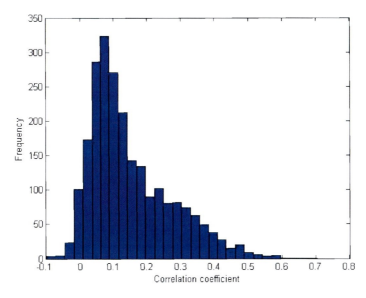

Fig. 1 Histogram of the correlation coefficients for the EG industry

3.1 Electricity, Gas and Water Supply Branch

There are 69 companies in this branch. Figure 1 shows histogram of the pairwise correlations between returns of stocks which compose the Electricity, Gas and Water Supply industry in the Russian market. Most of the correlation coefficients are concentrated in the interval $(0; 0.5)$. A lot of values belong to the interval $[0; 0.1]$. This fact means that there is free behaviour of companies inside this branch. We also observe approximately uniform decreasing of the histogram with increasing of correlation from value $r_0 = 0.15$.

Figure 2 shows dependence between the threshold values and edge density for the EG industry.

Note that the edge density changes sharply in the interval $[0; 0.4]$. Figure 2 allows to select thresholds for finding maximum cliques and independent sets. Intuitively, high positive thresholds are used for solving the maximum clique problem and nonpositive or small positive thresholds are used for solving the maximum independent set problem [1]. For example, values in the interval $[0.4; 1]$ are used as threshold for solving the maximum clique problem and values in the interval $[0; 0.1]$ are used as threshold for solving the maximum independent set problem.

Tables 1 and 2 show the results of finding the maximum clique and the maximum independent set of the market graph for a given threshold. Cliques and independent sets give us additional useful information concerning the behaviour of stocks in the branch [1–3]. Namely there are few stocks in the maximum independent set and clique. However according to histogram this is expected as for cliques, as for

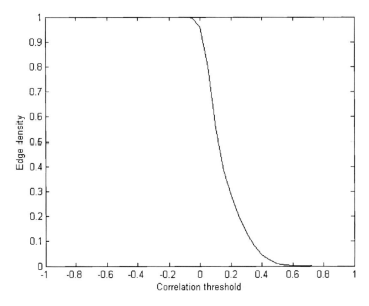

Fig. 2 Dependence between the threshold values and edge density of EG industry

Table 1 The maximum clique of EG branch in the Russian market for a given threshold

Threshold	Edge density	Size of clique	Companies in the maximum clique
0.5	0.0102	5	HYDR, MSNG, OGKA, OGKB, TGKA
0.45	0.0239	7	FEES, HYDR, MRKH, MRKHP, MSNG, OGKB, TGKA
0.4	0.0439	9	FEES, HYDR, MRKH, MRKHP, MRKY, MSNG, OGKA, OGKB, TGKA

independent sets because most of the correlation coefficients belong to the interval (0; 0.5). This means that a market graph constructed using correlation thresholds which are close to 0.5 will have small cliques and a market graph constructed using correlation thresholds which are close to 0 will have small independent sets.

Table 2 The maximum independent set of EG branch in the Russian market for a given threshold

Threshold	Edge density	Size of ind. set	Companies in the maximum ind. set
0	0.0384	3	YKEN, YKENP, KBSB
0.03	0.1172	5	LSNG, RZSB, VGSB, CLSBP, CLSBP
0.05	0.2012	7	LSNG, RZSB, TGKDP, TGKI, YKENP, CLSBP, CLSBP

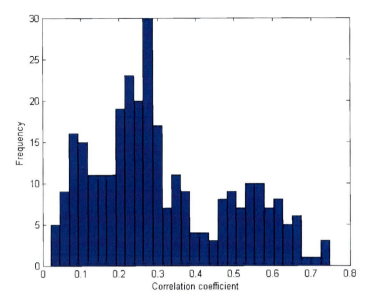

Fig. 3 Histogram of the correlation coefficients for the MQ branch

Table 3 The maximum clique of MQ branch in the Russian market for a given threshold

Threshold	Edge density	Size of clique	Companies in the maximum clique
0.65	0.0367	3	GAZP, ROSN, SNGS
0.6	0.0867	5	GAZP, ROSN, SNGS, TATN, UKUZ
0.55	0.14	7	GAZP, ROSN, SNGS, TATN, UKUZ, CHMF, MTLR

3.2 Mining and Quarrying Branch

There are 25 companies only. Figure 3 shows histogram of the pairwise correlations between returns of stocks which compose the MQ industry in the Russian market. This histogram differs from the histogram of EG industry. Notice two pronounced maximums in the histogram, in which the first maximum belongs to the interval [0.2; 0.3] and the second maximum belongs to the interval [0.5; 0.6]. Moreover, in the second interval there is the uniform distribution of correlations. This reflects a more varied structure of connections between stocks of MQ industry.

Figure 4 shows dependence between the threshold values and edge density for the EG industry.

Note that the edge density changes sharply in the interval [0.05; 0.65]. Figure 4 allows to select the thresholds for finding the maximum cliques and independent sets. For example, values in the interval [0.5; 1] are used as a threshold for solving the maximum clique problem and values in the interval [0.05; 0.25] are used as a threshold for solving the maximum independent set problem.

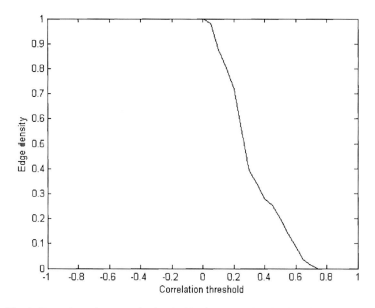

Fig. 4 Dependence between the threshold values and edge density of MQ industry

Tables 3 and 4 show the results of finding the maximum clique and the maximum independent set of market graph for a given threshold. Cliques and independent sets give us the same information concerning the behaviour of stocks in the branch as in previous branch.

Table 4 The maximum independent set of MQ branch in the Russian market for a given threshold

Threshold	Edge density	Size of ind. set	Companies in the maximum ind. set
0.07	0.05	3	MFGSP, LNZL, RUSP
0.09	0.09	4	CHEP, LNZL, PMTL, RUSP
0.12	0.15	6	SIBN, MFGSP, CHEP, LNZL, PMTL, RUSP

3.3 Chemicals and Chemical Products Branch

There are 11 companies. Figure 5 shows histogram of the pairwise correlations between returns of stocks which compose the CP industry in the Russian market. Most of the correlation coefficients are concentrated in the interval $(0.025, 0.25)$.

Analysis of Russian Industries in the Stock Market 61

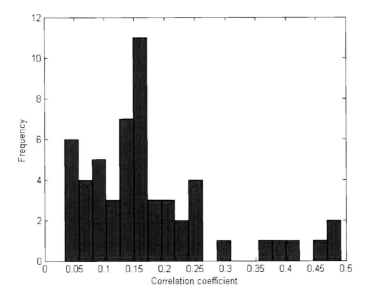

Fig. 5 Histogram of the correlation coefficients for the CP branch

This means that stocks in the CP branch are less dependent from each other compared with the MQ branch. The maximum correlation does not exceed 0.5.

Figure 6 shows dependence between the threshold values and edge density for the EG industry. Figure 6 allows to select the thresholds for finding the maximum clique and independent set. For example, values in the interval [0.3; 0.5] are used as a threshold for solving the maximum clique problem and values in the interval [0; 0.15] are used as a threshold for solving the problem of finding the maximum independent set.

Tables 5 and 6 show the results of finding the maximum clique and the maximum independent set of market graph with the same set of threshold values. The weak correlation between the returns of the stocks of the CP branch is confirmed by small size of cliques.

Consequently, the investigated characteristics don't allow to answer the question whether the stocks of each of the considered industries form clusters in market graph.

4 Cluster Analysis of the Russian Market

In this section we analyse the Russian market consisting of the considered branches as a whole. To check whether the stocks of each of the considered industries form clusters in a market graph we will distinguish inter/intra-edges of industries with high value of correlation. High value of correlation shows strongly correlated stocks.

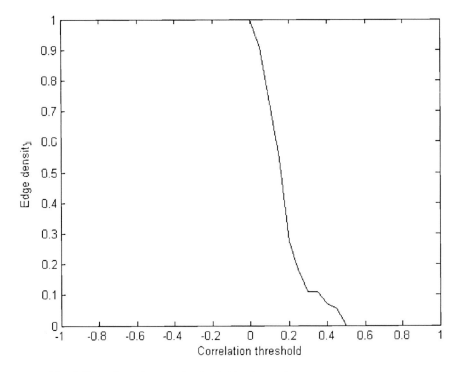

Fig. 6 Dependence between the threshold values and edge density of CP industry

The distribution of high correlations with respect to inter/intra-edges is presented in Table 7. In the first row there is the distribution of edges which connect stocks from different branches, for example, 5 inter-edges belonging to the interval [0.55; 0.65] connect branches. In the second row the number of intra-edges for each of all branches with weights from a specified range is given, for example, there are 13 intra-edges with the weights from the range of 0.65 to 0.75. And in the third row the number of edges with the weights from a specified range is given, for example, there are 146 edges in the Russian market with the weights from the range of 0.45 to 0.55.

Table 5 The maximum clique of CP branch in the Russian market for a given threshold

Threshold	Edge density	Size of clique	Companies in the maximum clique
0.4	0.0727	2	AKRN, DGBZ
0.3	0.1091	3	AKRN, DGBZ, DGBZP
0.2	0.2727	4	AKRN, DGBZ, DGBZP, URKA

Analysis of Russian Industries in the Stock Market 63

Table 6 The maximum independent set of CP branch in the Russian market for a given threshold

Threshold	Edge density	Size of ind. set	Companies in the maximum ind. set
0.05	0.0909	2	AZKM, KLNA
0.1	0.2727	3	AZKM, DGBZP, VLHZ
0.15	0.4545	5	AZKM, NKNCP, PHST, VLHZ, VRPH

Table 7 Number of high values of correlation coefficient of inter/intra-edges

	Values of correlation		
	[0.45, 0.55)	[0.55, 0.65)	[0.65, 0.75)
Inter	65	5	0
Intra	81	41	13
All	146	46	13

Table 8 Number of high values of correlation coefficient for three branches

	Values of correlation		
	[0.45, 0.55)	[0.55, 0.65)	[0.65, 0.75)
MQ	44	10	2
EG	34	31	11
CP	3	0	0

Table 9 Number of high values of correlation coefficient between branches

	Values of correlation		
	[0.45, 0.55)	[0.55, 0.65)	[0.65, 0.75)
EG and MQ	53	5	0
EG and CP	1	0	0
MQ and CP	11	0	0

The distributions of high correlations of stocks belonging to the same branch and to different branches are presented in Tables 8 and 9, respectively. The distribution of high correlations with respect to intra-edges of each of the branches separately is presented in Table 8. In the first row the number of intra-edges of MQ branch with the weights from a specified range is given, for example, there are 2 edges with the weights from the range of 0.65 to 0.75. In the second and the third rows there are the number of intra-edges of EG and CP industries with the weights from a specified range, respectively, for example, in the range of 0.55 to 0.65 the stocks of EG branch are connected by 31 edges and the stocks of CP branch don't have any connections. The number of inter-edges between each pair of branches is presented in Table 9. In the first row the number of inter-edges between EG and MQ branches with the weights from a specified range is given. In the second and the third rows

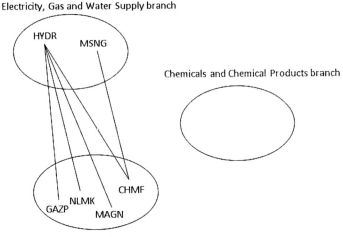

Fig. 7 Three industries with its intra-edges

Table 10 Degree of vertices in whole market

Threshold	Vertices					
	HYDR	MSNG	CHMF	MAGN	NLMK	GAZP
[0.55, 0.65)	7	4	11	7	10	11
[0.45, 0.55)	20	18	23	24	21	19

there are the number of inter-edges between EG and CP, MQ and CP branches with the weights from a specified range, respectively.

The first observation is that there are no connections between industries at the level higher than 0.65. The second observation is that there are 5 inter-edges only between branches MQ and EG with the weights from the range of 0.55 to 0.65. In order to analyse this observation deeper one can consider stocks which connected by this 5 inter-edges. Figure 7 shows these stocks with 5 inter-edges.

One can observe that MQ and EG industries are connected only via two stocks. This means that these branches will not be connected if stock HYDR and stock CHMF are deleted. Besides if stock HYDR and stock MSNG are deleted then considered MQ and EG industries will have no inter-edges between each other and with the CP industry too.

Table 10 shows the degrees of the vertices (companies) which are presented in Fig. 7. In the first and the second rows the degrees of these companies for two different ranges are given. Notice that all these vertices (companies) have a very high degree. Table 11 shows degrees of the same vertices only inside the corresponding industries.

This result shows that analysis of vertex degrees is also important for the overall market and market analysis by industry.

Table 11 Degree of vertices inside industries

Threshold	Vertices					
	HYDR	MSNG	CHMF	MAGN	NLMK	GAZP
[0.55, 0.65)	3	3	5	5	5	5
[0.45, 0.55)	10	8	6	6	5	5

5 Conclusion

Graph representation can be applied not only to the overall stock market but also to each of the industries in the market separately. Conventional approach to analyse the industries in the stock market such as studying histograms, finding cliques and independent sets give us additional information about the behaviour of the stocks in each industry. But this analysis does not provide sufficient information for the clustering of the market by the industry. However, we conduct a more detailed investigation which is related, in particular, with the degrees of vertices. This analysis could provide division of the market to some sets of the stocks which constitute industries in the stock market.

Acknowledgements The authors are partially supported by LATNA Laboratory, NRU HSE, RF Government Grant N. 11.G34.31.0057 and RFFI Grant 14-01-00807.

Appendix

Number	Code	Full names
1	BEGY	JSC Bashkirenergo common stock
2	DVEC	JSC Far East Energy Company common stock
3	EONR	JSC OGK 4 common stock
4	ETGK	JSC Yenisei TGK 13 common stock
5	FEES	JSC Federal Grid Company of Unified Energy System common stock
6	HYDR	JSC RusGidro
7	IRGZ	JSC Irkutskenergo common stock
8	KCHE	JSC Kamchatskenergo common stock
9	KISB	JSC Kirovenergosbyt common stock
10	KISBP	JSC Kirovenergosbyt preferred stock
11	KRSG	JSC Krasnoyarsk hydro electric station common stock
12	KUBE	JSC Energy and Electrification of Kuban
13	LSNG	JSC Energy and Electrification Lenenergo common stock

(continued)

Number	Code	Full names
14	LSNGP	JSC Energy and Electrification Lenenergo preferred stock
15	MRKC	JSC Interregional Distribution Grid Company of the center Caucasus common stock
16	MRKH	JSC Interregional Distribution Grid Companies Holding common stock
17	MRKHP	JSC Interregional Distribution Grid Companies Holding preferred stock
18	MRKK	JSC Interregional Distribution Grid Company of the North Caucasus common stock
19	MRKP	JSC Interregional Distribution Grid Company of the Center and Privolzhe common stock
20	MRKS	Interregional Distribution Grid Company of the Siberia common stock
21	MRKU	JSC Interregional Distribution Grid Company of Urals common stock
22	MRKV	Interregional Distribution Grid Company of Volgi common stock
23	MRKY	Interregional Distribution Grid Company of the South common stock
24	MRKZ	Interregional Distribution Grid Company of the North-West common stock
25	MSNG	Mosenergo common stock
26	MSRS	JSC Moscow United Electric Grid Company common stock
27	MSSB	JSC Mosenergosbyt common stock
28	MSSV	JSC Moscow United Energy Company common stock
29	OGKA	JSC OGK 1 common stock
30	OGKB	JSC OGK 2 common stock
31	OGKC	JSC OGK 3 common stock
32	OGKE	JSC OGK 5
33	PMSB	JSC Perm Energy Retail Company common stock
34	PMSBP	JSC Perm Energy Retail Company preferred stock
35	RKKE	Rocket and Space Corporation Energia Korolev common stock
36	RTSB	JSC Energosbyt Rostovenergo common stock
37	RTSBP	JSC Energosbyt Rostovenergo preferred stock
38	RZSB	JSC Ryazan Energy Retail Company common stock
39	STSBP	JSC Stavropolenergosbyt preferred stock
40	SVSB	JSC Sverdlovenergosbyt common stock
41	SVSBP	JSC Sverdlovenergosbyt preferred stock
42	TASB	JSC Tambov Energy Company common stock
43	TASBP	JSC Tambov Energy Company preferred stock
44	TGKA	TGK 1 common stock
45	TGKB	TGK 2 common stock
46	TGKBP	TGK 2 preferred stock
47	TGKD	JSC Kvadra common stock
48	TGKDP	JSC Kvadra preferred stock
49	TGKE	JSC TGK 5 common stock
50	TGKF	JSC TGK 6 common stock
51	TGKI	JSC TGK 9 common stock
52	TGKJ	JSC Fortum common stock
53	TGKK	JSC TGK 11 common stock

(continued)

Analysis of Russian Industries in the Stock Market

Number	Code	Full names
54	TGKN	JSC TGK 14 common stock
55	TORS	JSC Tomsk distribution company common stock
56	TORSP	JSC Tomsk distribution company preferred stock
57	UDSB	JSC Udmurt Energy Retail Company
58	UDSBP	JSC Udmurt Energy Retail Company preferred stock
59	VDSB	JSC Vladymirenergosbyt common stock
60	VGSB	JSC Volgograd Energy Retail Companycommon stock
61	VGSBP	JSC Volgograd Energy Retail Company preferred stock
62	VRAO	JSC Rao Energy System of East common stock
63	YKEN	JSC YakutskEnergo common stock
64	YKENP	JSC YakutskEnergo preferred stock
65	ASSB	JSC Astrakhan Energy Retail Company
66	KBSB	JSC Kubanenergosbyt common stock
67	CLSB	JSC Chelyabenergosbyt
68	CLSBP	JSC Chelyabenergosbyt preferred stock
69	DASB	Dagenstan Energy Retail Company common stock
70	AMEZ	JSC Ashin iron plant common stock
71	ARMD	JSC ARMADA common stock
72	CHEP	JSC Chelyabinsk Pipe Rolling Plant common stock
73	CHMF	JSC Severstal common stock
74	CHZN	Chelyabinsk Zinc Plant common stock
75	GMKN	JSC Norilsk Nickel common stock
76	IRKT	JSC Irkut Corporation common stock
77	LNZL	JSC Lenzoloto common stock
78	MAGN	JSC Magnitogorst Iron and Steel Plant common stock
79	MTLR	JSC Mechel common stock
80	NLMK	JSC Novolipetsk Metallurgical combine common stock
81	PLZL	JSC Polyus Gold
82	PMTL	JSC Polymetal common stock
83	RUSP	JSC Ruspolimet common stock
84	GAZP	JSC Gazprom common stock
85	ROSN	JSC Rosneft
86	LKOH	JSC Lukoil common stock
87	SIBN	JSC Gazprom Neft common stock
88	SNGS	JSC Surgutneftegas common stock
89	SNGSP	JSC Surgutneftegas preferred stock
90	TATN	JSC Tatneft common stock
91	TATNP	JSC Tatneft preferred stock
92	UKUZ	JSC South Kuzbass common stock
93	MFGS	Slavneft-Megionneftegaz common stock
94	MFGSP	Megionneftegas preferred stock
95	AKRN	JSC Akron common stock
96	AZKM	JSC Azot common stock
97	DGBZ	JSC Dorogobuzh common stock

(continued)

Number	Code	Full names
98	DGBZP	JSC Dorogobuzh preferred stock
99	KLNA	JSC Concern Kalina common stock
100	NKNC	JSC Nizhnekamskneftekhim common stock
101	NKNCP	JSC Nizhnekamskneftekhim preferred stock
102	PHST	JSC Pharmstandart common stock
103	URKA	JSC Uralkali common stock
104	VLHZ	JSC Vladimir Chemical Plant common stock
105	VRPH	JSC VEROPHARM common stock

References

1. Boginski, V., Butenko, S., Pardalos, P.M.: On structural properties of the market graph. In: Nagurney, A. (ed.) Innovations in Financial and Economic Networks, pp. 28–45 . Edward Elgar Publishers, Northampton (2003)
2. Boginski, V., Butenko, S., Pardalos, P.M.: Statistical analysis of financial networks. Comput. Stat. Data Anal. **48**, 431–443 (2005)
3. Boginski, V., Butenko, S., Pardalos, P.M.: Mining market data: a network approach. Comput. Oper. Res. **33**, 3171–3184 (2006)
4. Investment company Finam: Accessed 1 October 2013, <http://www.finam.ru/>
5. Vizgunov, A.N., Kalyagin, V.A., Koldanov, A.P.: Applying market graphs for Russian stock market analysis. J. New Econ. Assoc. **3**(15), 65–81 (2012)

A Particle Swarm Optimization Algorithm for the Multicast Routing Problem

Yannis Marinakis and Athanasios Migdalas

Abstract In this paper, a new algorithm for the solution of the Multicast Routing Problem based on Particle Swarm Optimization algorithm is presented and analyzed. A review of the most important evolutionary optimization algorithms for the solution of this problem is also given. Three different versions of the proposed algorithm are given and their quality is evaluated with experiments conducted on suitably modified benchmark instances of the Euclidean Traveling Salesman Problem from the TSP library. The results indicated the efficiency of the proposed method.

1 Introduction

Routing is the process of finding communication paths between nodes in order to transfer information from a source to its destinations. Routing uses a data structure called **routing table** at each node which at a minimum stores all neighboring nodes, i.e., nodes at one hop distance. However, it may store nodes farther away. In order to reach a specific destination, a router decides which neighboring node is selected from the routing table.

There are two main types of routing policies [43]: **static routing** and **dynamic** (or **adaptive**) **routing**. In static routing, the routes between nodes are static; they are precomputed and stored in the routing table. There is no communication between routers regarding the current topology of the network. In dynamic routing the routing table is not static; routes are generated based on prevailing factors such as

---

Y. Marinakis
Technical University of Crete, School of Production Engineering and Management, Decision Support Systems Laboratory, 73100 Chania, Greece
e-mail: marinakis@ergasya.tuc.gr

A. Migdalas (✉)
Department of Civil Engineering, Aristotle University of Thessalonike, 54124 Thessalonike, Greece

Industrial Logistics, Luleå Technical University, 97187 Luleå, Sweden
e-mail: samig@civil.auth.gr; athmig@ltu.se

bandwidth utilization and topology changes, for instance due to link failures. Thus, **adaptive routing algorithms** are devised in order to construct routing tables that make possible to forward communication packets based on information about the current network status.

Routing policies can be further divided into **centralized** or **distributed**. In the former case, a centralized node, possibly backed up by a few other nodes, is responsible to generate the routes for any pair of nodes. On the other hand, in distributed routing, every node generates routes locally and independently.

Routing protocols [13] specify how routers (switching nodes) communicate with each other in order to disseminate necessary information that enables the selection of routes between nodes. However, routing algorithms are responsible for the specific route choice. There exist several routing protocols which fall into two major categories, **distance-vector routing protocol** and **link-state protocol**. A distance-vector routing protocol requires that a router informs its neighbors of topology changes periodically. Its name refers to the fact that the protocol manipulates vectors of distances to other nodes in the network. On the other hand, link-state protocols require a router to inform all the nodes in a network of topology changes. Thus, distance-vector routing protocols have less computational complexity and message overhead. Examples of routing protocols are:

- The **Resource Information Protocol** (**RIP**) is a distance vector routing protocol that bases path selection on the metric of hop count.
- The **Interior Gate Protocol** (**IGP**) is another distance vector routing protocol which supports multiple metrics based on bandwidth, delay, load and Maximum Transfer Unit (MTU).
- The **Open Source Shortest Path First** (**OSPF**) is a link state routing protocol and probably the most common IGP which uses the shortest path first algorithm to compute low cost path to destination.

In this paper, a new algorithm for the multicast routing problem based on **Particle Swarm Optimization (PSO)** is proposed. **PSO** is a population-based swarm intelligence algorithm that was originally proposed by Kennedy and Eberhart [31] and simulates the social behavior of social organisms by using the physical movements of the individuals in the swarm. Its mechanism enhances and adapts to the global and local exploration. Most applications of PSO have concentrated on the optimization in continuous space while some work has been done on the discrete optimization [32, 45]. Recent surveys for Particle Swarm Optimization can be found in [4, 5, 9, 41]. Clerc and Kennedy [10] proposed a constriction factor in order to prevent explosion, ensure convergence and to eliminate the parameter that restricts the velocities of the particles.

The proposed algorithm is a hybridization of Particle Swarm Optimization (PSO), Variable Neighborhood Search (VNS) [25], and Local and Global Expanding Neighborhood Topology (LGENT). In the new approach:

- A particle is affected by the best local neighborhood when the number of neighbors is not equal to the whole swarm and by the best particle, otherwise.

- A different equation for the velocities is given where a fourth term has been added in the classic equation which represents the local neighbors.
- In the new equation the particle, except from a movement towards a new direction, a movement towards its previous best and a movement towards the global best of the swarm, can move towards the local best of its neighborhood.
- The neighborhood is expanding based on the quality of the solutions.

The rest of the paper is organized as follows: In the next section the Multicast Routing Problem is presented and analyzed in detail, while in Sect. 3 a review of the most important evolutionary algorithms for the problem is presented. The proposed algorithm is given and analyzed in Sect. 4, while in Sect. 5 the results and a discussion about the effectiveness of the proposed algorithm are presented. Finally, conclusions and future research directions are given in the last section.

2 The Multicast Routing Problem

Multicast routing is a technique to simultaneously transfer information from a source to a set of destinations in a communication network. Compared to **unicast routing** which is based on a point-to-point transmission, multicast is more efficient since it utilizes the inherent network parallelism, that is, shares resources, such as links and forwarding nodes, in order to efficiently deliver data from a source to destination nodes. Due to the increasing development of multimedia applications concerning video and audio transmissions but also the increased importance of teleconferencing, collaborative environments, distance learning, and e-commerce, the importance of multicast routing has increased dramatically and with it the interest of the scientific community in devising and developing efficient multicast routing algorithms [37, 38].

Most multicast routing algorithms have the goal of minimizing the cost of the constructed multicast tree. However, such a tree is a (**rooted**) **Steiner tree** and it is well known that the problem of finding it is NP-hard [24]. Therefore, much attention has been directed towards developing heuristics of polynomial complexity that produce near optimal results and often guarantee that the produced solutions are within twice the cost of the optimum one [8, 37, 38, 47, 49]. Moreover, the multicast routing has also been formulated as **constrained Steiner tree** problems. Thus, Kompleea et al. [33] require that each path in the tree must satisfy an end-to-end delay bound, Tode et al. [48] restrict the number of packet copies per network node, Hwang et al. [27] minimize the cost of the multicast tree subject to low path delay, and Wu and Hwang [50] consider the minimization of the cost of the tree subject to multiple constraints concerning such issues as end-to-end delay and end-to-end loss probability. The **multicast routing problem** (**MRP**) may be stated as follows:

Given a network $\mathscr{G} = (\mathscr{N}, \mathscr{E}, c)$ where $\mathscr{N} \neq \emptyset$ is the set of nodes, $\mathscr{E} \subset \mathscr{N} \times \mathscr{N}$ is the set of edges connecting pairs of nodes. Let $|\mathscr{N}| = n$ and $|\mathscr{E}| = m$. Moreover, let $s \in \mathcal{N}$ be a distinct node, called the source node, and let $\mathcal{D} \subseteq \mathcal{N}\setminus\{s\}$ be the set of destination nodes.

When a **Single Flow routing** problem is considered:

$$x_e^k = \begin{cases} 1 \text{ if link } e \text{ is used to transmit flow to destination } k, \\ 0 \text{ otherwise,} \end{cases} \forall k \in \mathcal{D}. \quad (1)$$

On the other hand, when a **Multi-Flow routing** is considered: Let $\mathcal{F} = \{1, 2, \ldots, \upsilon\}$ be the set of different flows emanating from the source s and let $\mathcal{D}_f \subseteq \mathcal{N}\setminus\{s\}$ for $f \in \mathcal{F}$ be the corresponding set of destination nodes, then:

$$x_e^{fk} = \begin{cases} 1 \text{ if link } e \text{ is used to transmit flow } f \text{ to destination } k, \\ 0 \text{ otherwise,} \end{cases} \forall k \in \mathcal{D}_f, \forall f \in \mathcal{F} \quad (2)$$

Consider the single objective problem flow case. Let $c : \mathcal{E} \to \mathcal{R}_0^+$ be a cost function that assigns a real value to each edge. We would like to find a subnetwork $\mathcal{T}(\{s\} \cup \mathcal{D}) = (\mathcal{N}_T, \mathcal{E}_T, c_T)$ of $\mathcal{G}$ such that:

- $\{s\} \cup \mathcal{D} \subseteq \mathcal{N}_T$.
- $\mathcal{E}_T \subset \mathcal{N}_T \times \subset N_T$.
- c_T is the restriction of c to $\mathcal{E}_T$.
- there is a path from s to every node $v \in \mathcal{D}$.
- the total cost of $\mathcal{T}(\{s\} \cup \mathcal{D})$, i.e. $c(\mathcal{T}(\{s\} \cup \mathcal{D})) = \sum_{e \in \mathcal{E}_T} c(e)$, is minimized.

The cost in this problem setting may refer to **various metrics** of network resource utilization such as delay, bandwidth, number of links, etc. The approach has received criticism because the minimization of the sum of the edge costs does not in general lead to the optimization of performance factors such as delay and bandwidth utilization [19].

The two most common and important requirements when designing multicast trees are **delay** and **bandwidth utilization**.

- A cost occurs from using and/or reserving network resources such as bandwidth.
- The **end-to-end delay** is the sum of the total delays encountered along the paths from source to each destination. Typically, the delay should be within a certain bound in real-time communications.

Such additional requirements must be imposed separately resulting in constrained versions of the problem [27, 36, 50]. Let $\mathcal{P}_v \subset \mathcal{E}_T$ denote, in the form of a sequence of edges, the path in $\mathcal{T}(\{s\} \cup \mathcal{D})$ from the source s to the destination $v \in \mathcal{D}$.

If besides c we, also, introduce the delay function $d : \mathscr{E} \to \mathscr{R}_0^+$, then, the delay $d(\mathscr{P}_v)$ along the path $\mathscr{P}_v$ is the sum of the delays on all links along the path, i.e. $d(\mathscr{P}_v) = \sum_{e \in \mathscr{P}_v} d(e)$.

The delay of the multicast tree $\mathscr{T}(\{s\} \cup \mathscr{D})$ is, then, defined as the maximum delay among all such paths, that is, $d(\mathscr{T}(\{s\} \cup \mathscr{D})) = \max_{v \in \mathscr{D}} \{d(\mathscr{P}_v)\}$.

Hence, the **delay-constrained least cost routing tree** must satisfy all the previous requirements plus the additional constraints:

$d(\mathscr{P}_v) \leq b$, $\forall v \in \mathscr{D}$, where b is a specified delay bound.

3 Evolutionary Algorithms for the Multicast Routing Problem

A number of evolutionary algorithms (based mainly on Genetic Algorithms and Ant Colony Optimization) have been proposed for solving the single objective and multiobjective routing problems. More precisely, Munetomo [35] derived a genetic routing algorithm called the **Genetic Adaptive Routing Algorithm (GARA)** which creates alternative routes in a routing table by applying genetic operators. It is a source routing algorithm, that is, it determines the entire route of a packet in its source node. At each node it maintains for each destination a limited number of alternative routes that are frequently used and which are determined by genetic operators. Each packet is assigned one such route to its destination. The fitness (weight) of each route specifies the probability of being selected among the existing alternatives. The algorithm calculates these weights based on the observed communication latency of the frequently used routes by sending delay query packets.

In GARA, the routing table consists of five columns: destination, route, frequency, delay, and weight (fitness). The frequency specifies the number of packets sent along the route to the destination, while the delay stores the communication latency of packets sent along the route. Initially, this routing table is empty. When necessity arises to sent a packet to a destination, Dijkstra's shortest path algorithm is used to determine a shortest path to the destination using the hop count metric. The communication latency of the route is observed by issuing a delay query message once the route has been assigned to a specific number of packets. In GARA, alternative routes are generated based on the topological information of the network by invoking the path mutation and path crossover genetic operators. The path mutation generates an alternative route by perturbing an existing one, while path crossover exchanges subroutes between pairs of routes. Selection operators are applied to prevent overflow of the routing table. One such operator deletes routes with lowest fitness, while another deletes routes with low utilization. In [35] computational comparisons to Routing Information Protocol (RIP) and adaptive Shortest-Path First (SPF) protocol for a network of 20 nodes are given in which GARA prevails.

Hwang et al. [27] develop a genetic algorithm for multicast routing using an encoding in which each gene X_k ($1 \leq k \leq |\mathscr{D}|$) of the chromosome $\mathbf{X}$ is an integer corresponding to a route $\mathscr{P}_{u_k}$ from the source s to the kth destination $u_k \in \mathscr{D}$ in the routing table. Thus, each chromosome guarantees a path between the source node and each destination node. However, the chromosome is not necessarily a tree and may contain cycles. The fitness of a chromosome $\mathbf{X}$ is calculated as $f(\mathbf{X}) = 1 - \dfrac{c(\mathbf{X})}{c(\mathscr{G})}$, where $c(\mathbf{X})$ is the sum of the costs on the links of the subgraph induced by $\mathbf{X}$ and $c(\mathscr{G})$ is the total cost of the network. Thus, $0 \leq f(\mathbf{X}) \leq 1$. The chromosomes in a genetic population (pool) are sorted in decreasing fitness and duplicated chromosomes are discarded by replacing each of them by a new randomly generated chromosome.

In the reproduction process, chromosomes with high fitness are selected to be introduced in the next generation and in order to produce offspring through crossover and mutation that will replace chromosomes with low fitness. The crossover operator randomly mates two chromosomes with high fitness. Mutation is applied in a pointwise fashion, that is, a bit is changed with a certain mutation probability. Computational results comparing the algorithm to a Steiner heuristic are given in [27].

Fabregat et al. [22] consider the multiobjective multicast routing problem with four objectives concerning the maximum link utilization, the hop count, the total bandwidth consumption, and the total end-to-end delay. Moreover, their proposal involves the concept of **multi-tree routing** in which flow from the source to the destination nodes is split into subflows each routed along different trees. Hybridization of genetic algorithms with other heuristics and metaheuristics are discussed by [1, 28, 29] in very general settings. Such an approach is followed by [51] who developed a highly involved hybrid algorithm based on simulated annealing and genetic local search for the multiobjective multicast routing problem with a number of objectives. The candidate multicast trees are encoded according to the proposal of [22], that is, each chromosome encodes a path to every destination. Each tree in the initial population is generated by starting from the source node and randomly selects the next neighboring node until all destination nodes are reached. During the entire process a set of obtained non-dominated solutions is maintained. In every population crossover and mutation operators are applied to randomly selected pairs of trees and local search is applied to the offspring. A crossover operator exchanges the paths of the parents between two points carefully in order to avoid the creation of cycles or the disconnection of the trees. The mutation operator uses an adaptive probability computed on the basis of the fitness of the individual and on the current temperature in order to alter the individual by randomly replacing a path by an alternative. Each newly generated solution is compared to its closest, in Euclidean distance, solution with respect to weighted scalarized objective values and either replaces it or is discarded. To tune the search direction when the search is getting close to Pareto front, the weight vector is adaptively tuned according to the closest non-dominated solution in the current population if the current temperature has decreased below a threshold. Xu et al. [51] perform extensive computations in order to test the impact of local search and the impact of adaptive mutation and, also, in order to compare with two other genetic approaches by [11, 12].

In a series of works, Dorigo, Di Caro, and associates [14–18, 20, 23, 44] present **AntNet** as an approach for routing in communication networks. In AntNet, they do retain the core ideas of the ant colony optimization algorithm but also perform the necessary modifications in order to match a distributed and dynamic context different from the typical optimization problem. AntNet is an adaptive, mobile-agents-based algorithm based on the ant colony algorithm and metaphor. In this metaphor, the ant colony algorithm, along with routing table and data packets, uses an intelligent packet, the ant, in order to find promising paths between nodes. Ant packets use the concept of **stigmergy**. Pinto et al. [39, 40] consider the multiobjective multicast routing problem with three objectives: (a) Tree cost, (b) Maximum end-to-end delay (c) Average delay. Their approach uses an ant colony for the construction of a population of routing trees in each generation. A set $\mathscr{F}$ of non-dominated trees are maintained and updated in order to produce the Pareto front.

Mukherjee and Acharyya [34] consider the ant colony optimization in order to route along the shortest path with optimum throughput. Three different versions of the basic approach are considered by [34]: (a) ants are allowed to create loops, (b) attempt is made to prohibit some loops by not allowing an ant to revisit the last node previously visited, and (c) ants are prohibited to revisit the last M nodes (tabu list) already visited. Cauvery and Viswanatha [7] attempt to hybridize the AntNet approach with a genetic algorithm. Ant multi-path routing is studied by Purkayastha [42] and two algorithms are investigated. Curran [13] adapts the ant colony algorithm as a learning strategy for ad-hoc routing learning problem. Kassabalidis et al. [30] and Arabshahi et al. [2] consider the adaption of AntNet to the case of adaptive routing in wireless networks. Sigel et al. [46] apply the ant colony optimization inspired by AntNet to adaptive routing of a satellite telecommunications network. Three variants of the algorithm are considered and their performances are compared to each other and to versions of SPF, OSPF, and Dijkstra's shortest path algorithm. Baras and Mehta [6] develop an algorithm inspired by AntNet in order to discover and maintain paths in a mobile ad-hoc network (MANET).

4 Particle Swarm Optimization with Combined Local and Global Expanding Neighborhood Topology (PSOLGENT)

In the PSO algorithm there are two kinds of population topologies: the global best (*gbest*) population topology and the local best (*lbest*) population topology. In the *gbest* PSO, the neighborhood for each particle is the entire swarm. The social network employed by the *gbest* PSO reflects the star topology in which all particles are interconnected. Thus, the velocities of each particle are updated based on the information obtained from the best particle of the whole swarm. In the *lbest* PSO, each particle has a smaller neighborhood. In this case, the network topology corresponds to the ring topology where each particle communicates with only a limited number of other members of the swarm. The communication is usually achieved using the particles' indices [21].

In this paper, two different alternative ways of neighborhood expansion are considered. In the first one, the neighborhood is expanding when the best particle has not improved for a specified number of iterations (it_{num}). In the second alternative, each particle has its own neighborhood. Thus, while all particles begin with the same neighborhood topology, when a solution of a particle has not improved for a consecutive number of iterations, then, its neighborhood alone is expanded. Thus, with this strategy, it is possible that different particles have different neighborhood topologies. This is another novelty of the proposed algorithm.

This idea of changing the neighborhood topology of the swarm not with the number of iterations but when the swarm cannot improve the global (as in the first alternative) or personal best solution (as in the second alternative) is inspired by the basic idea of the Variable Neighborhood Search (VNS) Algorithm [25] where neighborhoods are changed when the algorithm gets trapped in a local optimum. Thus, similar to the VNS algorithm, where the neighborhood is expanded in order to find a better local optimum, in the proposed algorithm the search for a better direction along which the particle will move is performed in an expanded particle neighborhood whenever the particle gets trapped in a local optimum. This incorporation of the basic VNS characteristic into the PSO results in a much more powerful version of the latter. Another characteristic of the VNS algorithm that is incorporated into the PSO is the reinitialization of the search of the local best neighborhood. That is, in the proposed algorithm, whenever the number of neighbors becomes equal to the number of particles, then, the search of the local best neighborhood is reinitialized from a very small neighborhood. In each iteration of the algorithm, the search is realized from a different point as the current position of each particle is different in each iteration (c.f. the shaking procedure of VNS).

A fourth term is added to the equation of velocities. It represents the interaction of each particle with its local best neighborhood. Thus, the proposed equation of velocities becomes (PSOLGENT):

$$v_{ij}(t+1) = \chi_1(v_{ij}(t) + c_1 rand_1(pbest_{ij} - x_{ij}(t))$$
$$+ c_2 rand_2(gbest_j - x_{ij}(t)) + c_3 rand_3(lbest_{ij} - x_{ij}(t))), \qquad (3)$$

where c_1, c_2, and c_3 are the acceleration coefficients, $rand_1$, $rand_2$, and $rand_3$ are three random parameters in the interval [0, 1] and

$$\chi_1 = \frac{2}{|2 - c - \sqrt{c^2 - 4c}|} \text{ and } c = c_1 + c_2 + c_3, c > 4, \qquad (4)$$

where t is the iterations' counter.

A particle's best position ($pbest_{ij}$) in a swarm is calculated from the equation:

$$pbest_{ij} = \begin{cases} x_{ij}(t+1), & \text{if } f(x_{ij}(t+1)) < f(x_{ij}(t)) \\ pbest_{ij}, & \text{otherwise.} \end{cases} \quad (5)$$

The optimal position of the whole swarm at time t is calculated from the equation:

$$gbest_j \in \{pbest_{1j}, pbest_{2j}, \cdots, pbest_{Nj}\} | f(gbest_j) = \min\{f(pbest_{1j}), f(pbest_{2j}), \cdots, f(pbest_{Nj})\}, \quad (6)$$

and the $lbest_{ij}$ in Eq. (3) is calculated as follows:

$$lbest_{ij} \in \{N_i | f(lbest_{ij}) = \min\{f(x_{ij})\}, \forall x \in N_i\} \quad (7)$$

where the neighborhood N_i is defined by:

$$N_i = \{pbest_{i-n_{N_i}}(t), pbest_{i-n_{N_i}+1}(t), \cdots, pbest_{i-1}(t), \\ pbest_i(t), pbest_{i+1}(t), \cdots, pbest_{i+n_{N_i}}(t)\}. \quad (8)$$

The position of a particle changes using the following equation:

$$x_{ij}(t+1) = x_{ij}(t) + v_{ij}(t+1). \quad (9)$$

4.1 Path Representation

All the solutions (particles) are represented by vectors of length equal to the number of nodes and of integer values $\{0, 1, 2, \cdots\}$, where a zero value means that the corresponding node does not belong to the path, the value 1 means that the node belongs to all paths, while the value 2 means that the node belongs only to the path number 1, the value 3 means that the node belongs only to the path number 2, and so on. For example, if we have a particle with ten nodes and starting node, the node 1, then a unicast routing with end node, the node 10, represented by the vector

1	2	3	4	5	6	7	8	9	10
1	0	0	1	1	0	1	1	0	1

means that the path is as follows:

$$1 \longrightarrow 4 \longrightarrow 5 \longrightarrow 7 \longrightarrow 8 \longrightarrow 10.$$

For a particle with ten nodes and starting node, the node 1, a multicast routing with two end nodes, the nodes 9 and 10, a vector representation of the form

$$\begin{array}{|cccccccccc|} \hline 1 & 2 & 3 & 4 & 5 & 6 & 7 & 8 & 9 & 10 \\ 1 & 0 & 2 & 1 & 3 & 0 & 2 & 0 & 2 & 3 \\ \hline \end{array}$$

means that the paths are as follows:

$$1 \longrightarrow 3 \longrightarrow 4 \longrightarrow 7 \longrightarrow 9$$

$$1 \longrightarrow 4 \longrightarrow 5 \longrightarrow 10.$$

4.2 Variable Neighborhood Search

A Variable Neighborhood Search (VNS) [25, 26] algorithm is applied in order to optimize the particles. Both combinatorial optimization local search (2−opt, 1 − 0 relocate, and 1 − 1 exchange) and continuous optimization local search algorithms are utilized. These are denoted by LS1,···, LS6.

All the algorithms are applied for a number of iterations (ls_{num}). The first one (LS1) uses a transformation of the solution inside the solution space. The second one (LS2) combines the current solution with the global best particle, while the third one (LS3) combines the current solution with the local best particle. The fourth one (LS4) produces a combination of the current solution, the global best and the local best particle. Finally, the fifth one (LS5) and the sixth one (LS6) perform crossover of the particle with the local best and the global best particle, respectively. The methods are described by the following equations:

$$\text{LS1: } x_{ij}(t_1 + 1) = rand_5 * x_{ij}(t_1), \tag{10}$$

$$\text{LS2: } x_{ij}(t_1 + 1) = rand_6 * gbest_j + (1 - rand_6) * x_{ij}(t_1), \tag{11}$$

$$\text{LS3: } x_{ij}(t_1 + 1) = rand_7 * lbest_{ij} + (1 - rand_7) * x_{ij}(t_1), \tag{12}$$

$$\text{LS4: } x_{ij}(t_1 + 1) = rand_8 * rand_9 * gbest_j$$
$$+ rand_8 * (1 - rand_9) * lbest_{ij} + (1 - rand_8) * x_{ij}(t_1), \tag{13}$$

$$\text{LS5: } x_{ij}(t_1 + 1) = \begin{cases} lbest_{ij}, & \text{if } rand_{10} \leq 0.5 \\ x_{ij}(t_1), & \text{otherwise,} \end{cases} \tag{14}$$

$$\text{LS6: } x_{ij}(t_1 + 1) = \begin{cases} gbest_j, & \text{if } rand_{11} \leq 0.5 \\ x_{ij}(t_1), & \text{otherwise,} \end{cases} \tag{15}$$

where t_1 is the local search iteration number, $rand_6, \cdots, rand_{11}$ are random numbers in the interval [0,1] and $rand_5$ is a random number in the interval [−1,1].

PSO for the MRP

Table 1 Parameters for all versions of the algorithm

	PSOLGNT	PSOLGENT1	PSOLGENT2
Particles	100	100	100
Iterations	1000	1000	1000
ls_{num}	10	10	10
it_{num}	–	10	10
Neighbors	5	3 to 99	3 to 99
c_1	1.35	1.35	1.35
c_2	1.35	1.35	1.35
c_3	1.40	1.40	1.40

Two different versions of the VNS algorithm are considered. The first one is the classic version. It starts with a certain neighborhood and when a local optimum is found w.r.t. to that one, the algorithm proceeds with the next (enlarged) neighborhood in turn. In the second version, called sequential VNS (SVNS), all selected neighborhoods are applied sequentially in each iteration. These algorithms are utilized both with continuous neighborhoods (VNS1 and SVNS1) and with discrete ones (VNS2 and SVNS2).

5 Results and Discussion

The algorithm was implemented in modern Fortran and tested on five modified benchmark instances from the TSPLIB: Eil51, Eil76, pr264, A280, and pr439. The TSP instances were modified to suitable instances of the multicast routing problem with the number of nodes ranging from 51 to 439, a single source (the first node) and two, three, or five destination nodes corresponding to highest indexed nodes (the last nodes). There is no connection between the destination nodes and the links between the other nodes are uni-directional. The link cost is the distance between the two end nodes of the cost, while the link delay is chosen randomly. Finally, a different delay bound b is associated with each instance. The value of b depends on the number of nodes and the random delay values.

The parameters of the proposed algorithm were selected after thorough testing of a number of alternative values. Those which gave the best results w.r.t. the solution quality and the computational time are listed in Table 1.

The results given in Table 2 are for three versions of PSO that do not employ local search (WLS) and four versions of PSO based on VNS: VNS1, VNS2, SVNS1, and SVNS2.

In all tables and figures of this section, PSOLGNT denotes the variant of the proposed algorithm which uses the new equation of velocities but a static number of local neighborhoods, PSOLGENT1 denotes the version which employs two new features, that is, the new equation of velocities and the expanding neighborhood procedure based on the best particle, while PSOLGENT2 is the variant with the expanding neighborhood procedure based instead on the personal best position of each particle.

Table 2 Results of the algorithm versions for all benchmark instances

	2 end nodes			3 end nodes			5 end nodes		
	PSOLGNT	PSOLGENT1	PSOLGENT2	PSOLGNT	PSOLGENT1	PSOLGENT2	PSOLGNT	PSOLGENT1	PSOLGENT2
eil51									
WLS	1138.33	1134.48	1111.48	1156.55	1173.3	1195.24	1198.88	1168.61	1192.67
VNS1	714.86	620.2	609.68	716.81	766.61	787.13	624.53	663.89	746.67
VNS2	40.32	40.32	40.32	70.93	66.99	72.95	217.68	228.38	263.41
SVNS1	514.93	494.6	480.86	486.91	529.3	523.66	436.05	585.88	575.8
SVNS2	40.32	40.32	40.32	184.04	158.1	298.49	301.92	347.92	341.17
eil76									
WLS	2045.34	2070.068	1978.26	2081.621	1965.782	1819.003	1962.035	1932.189	1900.051
VNS1	1286.936	1270.507	1287.74	1357.523	1333.8	1313.869	1257.165	1568.579	1460.93
VNS2	49.26	49.68	58.27	100.76	224.78	184.67	350.2694	372.89	297.003
SVNS1	982.36	967.59	1025.45	883.63	889.59	886.02	839.79	821.76	859.54
SVNS2	75.16	72.31	113.51	351.18	400.24	402.32	690.67	554.06	519.18
pr264									
WLS	108571.8	111604.2	107558.2	120118.6	125187.9	120216.8	154892.8	156176.4	183879.7
VNS1	75691.6	64693.97	66538.3	80139.88	75062.42	74962.39	99901.38	97441.84	97458.36
VNS2	40796.37	23029.15	25133.66	59726.99	56202.67	59568.67	67859.46	73662.71	73985.34
SVNS1	67842.44	70433.64	71303.2	82955.7	79964.22	81332.49	93282.33	97572.02	97342.74
SVNS2	63308.6	48494.66	44260.11	78409.79	67230.2	72943.94	92508.23	91243.62	92957.12

	2 end nodes			3 end nodes			5 end nodes		
	PSOLGNT	PSOLGENT1	PSOLGENT2	PSOLGNT	PSOLGENT1	PSOLGENT2	PSOLGNT	PSOLGENT1	PSOLGENT2
	A280								
WLS	4291.98	4470.234	3830.7	4613.914	4856.828	4609.352	6124.666	6064.632	6436.14
VNS1	4126.981	4146.048	4167.342	2888.697	2866.077	2888.345	3934.386	3396.083	3777.706
VNS2	1508.222	1423.462	1391.016	1466.969	1573.112	1413.364	2222.277	2687.921	2530.963
SVNS1	2772.248	3016.28	2827.097	3156.021	3152.754	3153.149	3130.198	3094.094	3149.1
SVNS2	1542.632	1221.802	1444.067	2194.929	2059.229	2087.406	2473.799	2518.661	2557.19
	pr439								
WLS	335412.5	292549.1	352364.2	421655.5	379063.6	410003	483956.7	552124.5	471173.5
VNS1	290356.9	167689.7	174279.3	176416.7	301053.1	198538	244557.2	482690.6	200821.6
VNS2	76115.86	73313.09	73458.76	73632.12	96066.55	79018	104698.6	135617.2	134594.3
SVNS1	166381.6	142149.6	137240.6	177385.7	170619.3	178557	194711.4	206554.3	200500.3
SVNS2	142266.1	105677	99727.28	147354.8	157639.9	115567.7	159415.4	176267	179654.2

All figures of this section show the improvement progress of the solution associated with the best particle during the iterations.

The first block of Table 2 presents 45 different run sets each containing 5 runs. Each set is distinguished by the variant of the PSO algorithm and the number of destination nodes in the instance (c.f. the columns of the table). It can be observed that for all sets of runs the variants that do not employ local search perform worse than all the other variants of PSO. It can be also noted that the versions of PSO with VNS2 and SVNS2 perform better than those employing VNS1 and SVNS1. Moreover, VNS2 performs in most cases better than SVNS2. In total, VNS2 performs best in 40 of the run sets, SVNS2 performs best in 2 sets while in the remaining three run sets, VNS2 and SVNS2 have equal performance.

However, comparing PSOLGNT to PSOLGENT1 and PSOLGENT2, it is clear that none variant dominates the others for all problem instances. A number of 75 run sets were created using the three different variants of PSO employing the same local search procedure for instances with the same number of destination nodes (cf. the rows of Table 2). More precisely, when no local search is employed, PSOLGNT performs best in 3 run sets, PSOLGENT1 in 4, and PSOLGENT2 in 8. When VNS1 is used, PSOLGNT performs best in 5 run sets, PSOLGENT1 in 6, and PSOLGENT2 in 4. If VNS2 is used, then all variants find the same solution for one instance, PSOLGNT performs best in 7 run sets, PSOLGENT1 in 4, and PSOLGENT2 in 3. When SVNS1 is used, PSOLGNT performs best in 7 run sets, PSOLGENT1 in 6, and the PSOLGENT2 in 2. Finally, in the case of SVNS2, all variants find the same solution for one instance, PSOLGNT performs best in 5 run sets, PSOLGENT1 in 6, PSOLGENT2 in 3. All in all, in two cases all the variants find the same solutions, PSOLGNT performs best in 27 run sets, PSOLGENT1 in 26, and PSOLGENT2 in 20.

In Figs. 1, 2, 3, 4, 5, and 6, a graphical presentation is given for 6 out of the first 45 run sets discussed previously. In these figures, six representative run sets

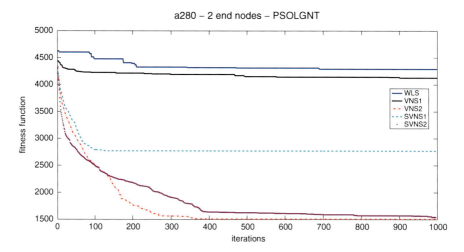

Fig. 1 Performance comparison of WLS, all VNS variants, and PSOLGNT on instance a280 with 2 destination nodes

PSO for the MRP   83

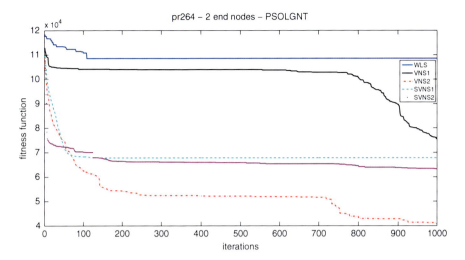

Fig. 2 Performance comparison of WLS, all VNS variants, and PSOLGNT on instance pr264 with 2 destination nodes

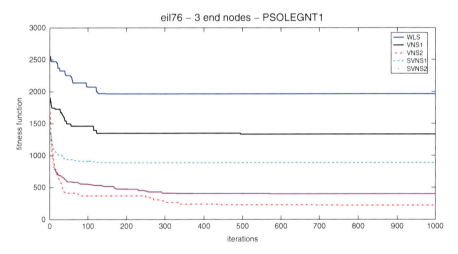

Fig. 3 Performance comparison of WLS, all VNS variants, and PSOLGNT on instance eil76 with 3 destination nodes

are depicted, taking care to show runs for all instances and for different number of destination nodes every time. Thus, in Figs. 1, 2, 3, 4, 5, and 6 the reduction of the objective function of the best particle is demonstrated for runs (a) of PSOLGNT without employing any local search algorithm, and with all VNS variants for the instance a280 with 2 destination nodes, (b) of PSOLGNT for the instance pr264 with 3 destination nodes, (c) of PSOLGENT1 for the instance eil76 with 5 destination nodes, (d) of PSOLGENT1 for the instance pr439 with 5 destination

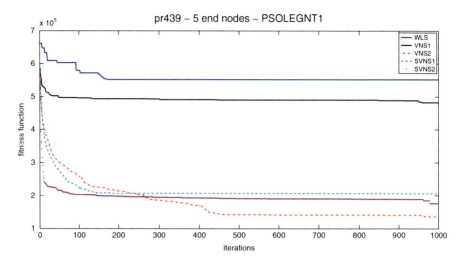

Fig. 4 Performance comparison of WLS, all VNS variants, and PSOLGNT1 on instance pr439 with 5 destination nodes

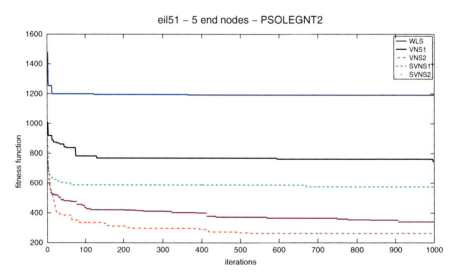

Fig. 5 Performance comparison of WLS, all VNS variants, and PSOLGNT2 on instance eil51 with 5 destination nodes

nodes, (e) of PSOLGENT2 for the instance eil51 with 5 destination nodes, and, finally, (f) of PSOLGENT2 for the instance a280 also with 5 destination nodes.

It can be seen on all figures that when no local search algorithm is utilized the results are less favorable than those obtained by variants PSO that employ VNS. Moreover, the results obtained when VNS1 is utilized are always worse than those obtained when other variants of VNS are employed. The results by SVNS1 are also

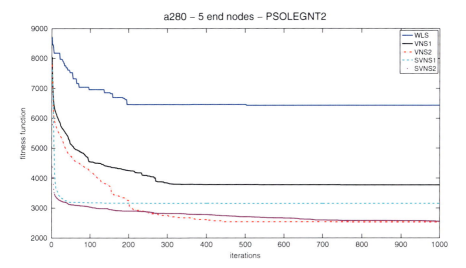

Fig. 6 Performance comparison of WLS, all VNS variants, and PSOLGNT2 on instance a280 with 5 destination nodes

worse than those obtained by employing either VNS2 or SVNS2. In some instances, the reduction in the objective function obtained by SVNS1 is almost equal to that obtained by the other two variants, however, as iterations progress, the process gets trapped in some local optimum and does not succeed to escape. In Fig. 4, the results obtained by employing SVNS1 are comparative with those obtained by SVNS2. The two best performing variants of VNS are VNS2 and SVNS2. However, the VNS2 converges faster and to better values than SVNS2.

In Figs. 7, 8, 9, 10, 11, and 12, 6 of the second 75 run sets are depicted graphically. In these figures, only the results obtained using VNS2 and SVNS2 are presented since the other two variants of VNS are inferior as the previous discussion has shown. Hence, in Figs. 7, 8, 9, 10, 11, and 12, the reduction of the objective function value for the best particle of all variants of PSO employing VNS2 and SVNS2 is shown for the problem instance eil51 with 5 destination nodes, for pr264 with 3 destination nodes, for a280 with 2 destination nodes, for pr439 with 2 destination nodes, for eil76 with 5 destination nodes, and, finally, for the instance pr264 with 2 destination nodes. From these results it is obvious that no variant performs better than the others for all problem instances. It can be seen, however, that when the VNS2 variant is employed, the convergence is faster and better objective function values are found regardless of the PSO variant used. We may conclude that the two versions of the PSO algorithm, differentiated by the type of the expanding neighborhood procedure, PSOLGENT1 and PSOLGENT2, and the variant with static number of local neighborhoods, PSOLGENT, perform equally well as long as they all employ the extended velocity equation and the same local search procedure.

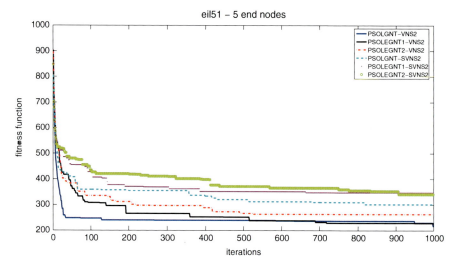

Fig. 7 Results of the three variants of PSO with VNS2 and SVNS2 using 5 end nodes for the instance eil51

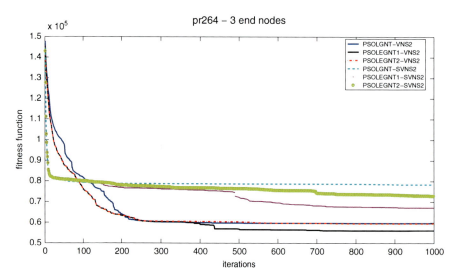

Fig. 8 Results of the three variants of PSO with VNS2 and SVNS2 using 3 end nodes for the instance pr264

6 Conclusions

In this paper, a new hybridized version of the Particle Swarm Optimization algorithm with the Variable Neighborhood Search was presented for solving the Multicast Routing Problem. The proposed algorithm combines a Particle Swarm

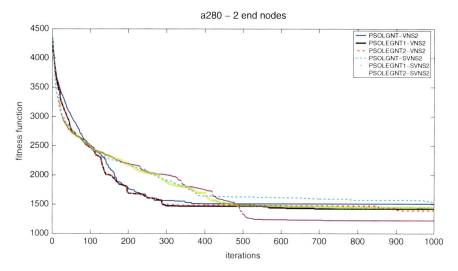

Fig. 9 Results of the three variants of PSO with VNS2 and SVNS2 using 2 end nodes for the instance a280

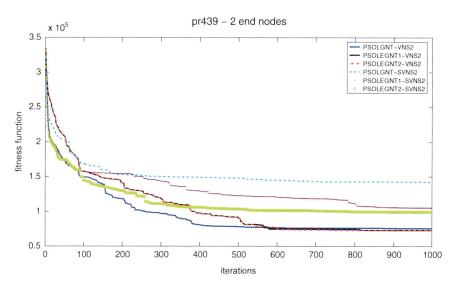

Fig. 10 Results of the three variants of PSO with VNS2 and SVNS2 using 2 end nodes for the instance pr439

Optimization with Local and Global Expanding Neighborhood Topology (PSOL-GENT) and extended with a fourth term velocity equation. In the new equation, the particle determines its movement along a new direction by combining a movement towards its previous best, a movement towards the global best of the swarm, and a movement towards the local best of its neighborhood. Moreover, a novel

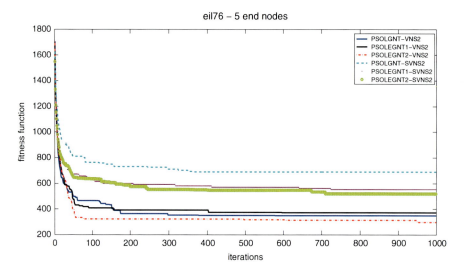

Fig. 11 Results of the three variants of PSO with VNS2 and SVNS2 using 5 end nodes for the instance eil76

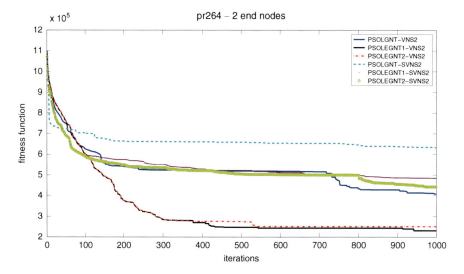

Fig. 12 Results of the three variants of PSO with VNS2 and SVNS2 using 2 end nodes for the instance pr264

expanding neighborhood topology has been introduced. The proposed algorithm expands neighborhoods on the basis of solution quality. We have introduced two different alternatives of neighborhood expansion. In the first one, the neighborhood is expanded when the best particle does not improve during a specified number of iterations. In the second alternative, each particle has its own neighborhood and when its associated solution does not improve for a number of consecutive iterations, only its own neighborhood is expanded. Hence, the VNS algorithm is applied in order to optimize the particles' position. Several variants of the algorithm were tested on a number of modified instances from the TSPLIB and the importance of employing VNS was verified. Future research will focus on improving the implementation and on the application of the methodology to other difficult problems.

References

1. Aarts, E.H.L., Verhoeven, M.G.A.: Genetic local search for the traveling salesman problem. In [3], chapter G9.5, G9.5:1–G9.5:7
2. Arabshahi, P., Gray, A., Kassabalidis, I., Das, A.: Adaptive routing in wireless communication networks using swarm intelligence. In: Proceedings of the 19th AIAA Int. Commun. Satellite Syst. Conf. (2001)
3. Bäck, T., Fogel, D.B., Michalewicz, Z. (Eds.): Handbook of Evolutionary Computation. Oxford University Press, Oxford (1997)
4. Banks, A., Vincent, J., Anyakoha, C.: A review of particle swarm optimization. Part I: background and development. Nat. Comput. **6**(4), 467–484 (2007)
5. Banks, A., Vincent, J., Anyakoha, C.: A review of particle swarm optimization. Part II: hybridisation, combinatorial, multicriteria and constrained optimization, and indicative applications. Nat. Comput. **7**, 109–124 (2008)
6. Baras, J.S., Mehta, H.: A probabilistic emergent routing algorithm for mobile ad hoc networks. In: WiOpt'03: Modeling and Optimization in Mobile, AdHoc and Wireless Networks, Sophia-Antipoli, France, INRIA, March 3–5 2003
7. Cauvery, N.K., Viswanatha, K.V.: Routing in dynamic network using ants and genetic algorithm. Int. J. Comput. Sci. Network Secur. **9**(3), 194–200 (2009)
8. Chow, C.H.: On multicast path finding algorithms. In: IEEE INFOCOM'91, pp. 1974–1283. IEEE, San Francisco (1991)
9. Clerc, M.: Particle Swarm Optimization. ISTE, London (2006)
10. Clerc, M., Kennedy, J.: The particle swarm: explosion, stability and convergence in a multidimensional complex space. IEEE Trans. Evol. Comput. **6**, 58–73 (2002)
11. Crichigno, J., Baran, B.: Multiobjective multicast routing algorithm for traffic engineering. In: Proceedings of the 13th International Conference on Computer Communication Networks, CCCN 2004, pp. 301–306. IEEE, San Francisco (2004)
12. Crichigno, J., Baran, B.: Multiobjective multicast routing algorithm. In: Lorenz, P., de Souza, J.N., Dini, P. (eds.), Telecommunications and Networking - ICT 2004. 11th International Conference on Telecommunications, Fortaleza, Brazil, August 1–6, 2004. Proceedings, vol. 3124 of Lecture Notes in Computer Science, pp. 1029–1034. Springer, New York (2004)
13. Curran, E.: Swarm: Cooperative reinforcement learning for routing in ad-hoc networks. Master's thesis, University of Dublin, Trinity College, September (2003)
14. Di Caro, G.: Ant colony optimization and its application to adaptive routing in telecommunication networks. PhD thesis, Université Libre de Bruxelles, Faculté des Sciences Appliquées, September (2004)
15. Di Caro, G., Dorigo, M.: Antnet: Distributed stigmergetic control for communication networks. J. Artif. Intell. Res. **9**, 317–365 (1998)
16. Di Caro, G., Dorigo, M.: Two ant colony algorithms for best-effort routing in datagram network. In: Proceedings of the 10th IASTED International Conference on Parallel and Distributed Computing and Systems (PDCS), (1998)

17. Di Caro, G., Dorigo, M.: Ant colonies for adaptive routing in packet-switched communications networks. In Eiben, A.E., Bäck, Th., Schoenauer, M., Schwefel, H.-P. (eds.), Parallel Problem Solving from Nature. PPSN V, vol. 1498 of Lecture Notes in Computer Science, pp. 673–682. Springer, New York (1998)
18. Di Caro, G.A., Ducatelle, F., Gambardella, L.M.: Theory and practice of ant colony optimization for routing in dynamic telecommunications networks. In: Sala, N., Orsucci, F. (eds.), Reflecting Interfaces: The Complex Coevolution of Information Technology Ecosystems, pp. 185–216. Idea Group, Hershey (2008)
19. Doar, M., Leslie, I.: How bad is naive multicast routing. In: INFOCOM'93. Proceedings. Twelfth Annual Joint Conference of the IEEE Computer and Communications Societies. Networking: Foundation for the Future, pp. 82–89. IEEE, San Francisco (1993)
20. Ducatelle, F., Di Caro, G., Gambarella, L.M.: Principles and applications of swarm intelligence for adaptive routing in telecommunications networks. Swarm Intell. **4**(3), 173–198 (2010)
21. Engelbrecht, A.P.: Computational Intelligence: An Introduction, 2nd edn. Wiley, Chichester (2007)
22. Fabregat, R., Donoso, Y., Solano, F., Marzo, J.L.: Multitree routing for multicast flows: A genetic algorithm approach. In: Vitriá, J., Radeva, P., Aguiló, I. (eds.), Recent Advances in Artificial Intelligence Research and Development, pp. 399–405. IOS Press, Amsterdam (2004)
23. Farooq, M., Di Caro, G.: Routing protocols for next generation networks inspired by collective behaviors of insect societies: An overview. In: Blum, C., Merkle, D. (eds.), Swarm Intelligence: Introduction and Applications, pp. 101–160. Springer, New York (2008)
24. Garey, M.R., Johnson, D.S.: Computers and Intractability: A Guide to the Theory of NP-Completeness. W. H. Freeman and Company, New York (1979)
25. Hansen, P., Mladenović, N.: Variable neighborhood search: Principles and applications. Eur. J. Oper. Res. **130**, 449–467 (2001)
26. Hansen, P., Mladenović, N., Moreno-Pérez, J.A.: Variable neighbourhood search: methods and applications. Ann. Oper. Res. **175**, 367–407 (2010)
27. Hwang, R.-H., Do, W.-Y., Yang, S.-C.: Multicast routing based on genetic algorithms. In: WiOpt'03: Modeling and Optimization in Mobile, Ad Hoc and Wireless Networks. INRIA Sophia-Antipolis, France, March 3–5, 2003
28. Ibaraki, T.: Combination with local search. In: [3], chapter G3.2, D3.2:1–D3.2:5
29. Ibaraki, T.: Simulated annealing and tabu search. In: [3], chapter D3.5, D3.5:1–D3.5:2
30. Kassabalidis, I., El-Sharkawi, M.A., Marks II, R.J., Arabshahi, P., Gray, A.A.: Swarm intelligence for routing in communication networks. In: Proceedings of the IEEE Globecom 2001, San Antonio, Texas (2001).
31. Kennedy, J., Eberhart, R.: Particle swarm optimization. In: Proceedings of 1995 IEEE International Conference on Neural Networks, vol. 4, pp. 1942–1948 (1995)
32. Kennedy, J., Eberhart, R.: A discrete binary version of the particle swarm algorithm, In: Proceedings of 1997 IEEE International Conference on Systems, Man, and Cybernetics, vol. 5, pp. 4104–4108 (1997)
33. Kompleea, V.P., Pasquale, J.C., Polyzos, G.C.: Multicast routing for multimedia communication. IEEE/ACM Trans. Network. **1**(3), 286–292 (1993)
34. Mukherjee, D., Acharyya, S.: Ant colony optimization techniques applied in network routing problem. Int. J. Comput. Appl. **1**(15), 66–73 (2010)
35. Munetomo, M.: The genetic adaptive routing algorithm. In: Corne, D.W., Oates, M.J., Smith, G.D. (eds.), Telecommunications Optimization: Heuristic and Adaptive Techniques, pp. 151–166. Wiley, Chichester (2000)
36. Oh, J., Pyo, I., Pedram, M.: Constructing minimal spanning/steiner trees with bounded path length. In: European Design and Test Conference, pp. 244–249 (1996)
37. Oliveira, C.A.S., Pardalos, P.M.: A survey of combinatorial optimization problems in multicast routing. Comput. Oper. Res. **32**(8), 1953–1981 (2005)
38. Oliveira, C.A.S., Pardalos, P.M., Resende, M.G.C.: Optimization problems in multicast tree construction. In: [43], 701–731

39. Pinto, D., Barán, B.: Multiobjective multicast routing with ant colony optimization. In: Gaiti, D. (ed.), Network Control and Engineering for QoS, Security and Mobility V, vol. 213 of IFIP International Federation of Information Processing, pp. 101–115. Springer, New York (2006)
40. Pinto, D., Barán, B., Fabregat, R.: Multi-objective multicast routing based on ant colony optimization. In: López, B., Meléndez, J., Radeva, P., Vitriá, J. (eds), Artificial Intelligence Research and Development, pp. 363–370. IOS Press, Amsterdam (2005)
41. Poli, R., Kennedy, J., Blackwell, T.: Particle swarm optimization. an overview. Swarm Intell. **1**, 33–57 (2007)
42. Purkayastha, P.: Multipath routing algorithms for communication networks: ant routing and optimization based approaches. PhD thesis, Department of Electrical and Computer Engineering, University of Maryland (2009)
43. Resende, M.G.C., Pardalos, P.M. (eds.): Handbook of Optimization in Telecommunications. Springer, New York (2006)
44. Saleem, M., Di Caro, G.A., Farooq, M.: Swarm intelligence based routing protocol for wireless sensor networks: Survey and future directions. Inform. Sci. **181**, 4597–4624 (2011)
45. Shi, Y., Eberhart, R.: A modified particle swarm optimizer. In: Proceedings of 1998 IEEE World Congress on Computational Intelligence, pp. 69–73 (1998)
46. Sigel, E., Denby, B., Le Hégarat-Mascle, S.: Application of ant colony optimization to adaptive routing in a leo telecommunications satellite network. Annales des Télecommunications **57**(5–6), 520–539 (2002)
47. Takahashi, H., Mutsuyama, A.: An approximate solution for the steiner problem in graphs. Mathematica Japonica **6**, 573–577 (1980)
48. Tode, H., Sakai, Y., Yamamoto, M., Okada, H., Tezuka, Y.: Multicast routing algorithm for nodal load balancing. In: IEEE INFOCOM'92, pp. 2086–2095. IEEE, San Francisco (1992)
49. Waxman, B.M.: Routing of multipoint connections. IEEE J. Sel. Area Comm. **1**(3), 286–292 (1988)
50. Wu, J.J., Hwang, R.-H.: Multicast routing with multiple constraints. Inform. Sci. **124**, 29–57 (2000)
51. Xu, Y., Qu, R., Li, R.: A simulated annealing based genetic local search algorithm for multi-objective multicast routing problems. Ann. Oper. Res. 1–29 (2013)

König Graphs for 4-Paths

Dmitry Mokeev

Abstract We give characterization of the graphs, whose each induced subgraph has the property: the maximum number of induced 4-paths is equal to the minimum cardinality of the set of vertices such as every induced 4-path contains at least one of them.

1 Introduction

Let $\mathscr{X}$ be a set of graphs. A set of disjoint induced subgraphs of graph G isomorphic to graphs in $\mathscr{X}$ is called $\mathscr{X}$-packing of G. A Problem of graph packing is to find maximum one in a graph. A subset of vertices of graph G which covers all induced subgraphs of G isomorphic to graphs in $\mathscr{X}$ is its $\mathscr{X}$-covering. A Problem of graph covering is to find minimum one in a graph. A König graph for $\mathscr{X}$ [1] is a graph in which every induced subgraph has the property that the maximum cardinality of its $\mathscr{X}$-packing is equal to the maximum cardinality of its $\mathscr{X}$-covering. The class of all König graphs for set $\mathscr{X}$ is denoted as $\mathscr{K}(\mathscr{X})$. If $\mathscr{X}$ consists of a single graph H, then we will talk about H-packings, H-coverings and König graphs for H.

One can find some similar terms in the literature: "König-Egervary graph" [2], "a graph with the König property" [6], "König graph" [7]. They all have the same meaning which is graph, in which cardinalities of maximum matching and minimum vertex cover are equal. Note that the definition of a König graph in this paper is not a generalization of the concept. This definition and usage of the term is motivated by the fact that the class of bipartite graphs, referred to in König Theorem, is exactly the class of all graphs whose cardinalities of maximum matching and minimum vertex cover are equal not only for a graph but also for all its induced subgraphs. Thus, the class of bipartite graphs coincides with the class of all König graphs for P_2 in this sense

---

D. Mokeev (✉)
Laboratory of Algorithms and Technologies for Networks Analysis, National Research University Higher School of Economics, 136, Rodionova Str., N.Novgorod, Russia
e-mail: MokeevDB@gmail.com

A lot of papers on the Problem of graph packing are devoted, especially on its algorithmic aspects (see, e.g., [5, 8]). It is known that the Problem of H-packing is NP-complete for any graph H, having a connected component with three or more vertices.

Being formulated as integer linear programming problem, the problems of $\mathscr{X}$-packing and $\mathscr{X}$-covering form a pair of dual problems. So König graphs are graphs such that for any induced subgraph there is no duality gap. In this regard König graphs are similar to perfect graphs having the same property with respect to another pair of dual problems (vertex coloring and maximum clique), which helps to solve efficiently these problems on perfect graphs [4].

Class $\mathscr{K}(\mathscr{X})$ is hereditary for any $\mathscr{X}$ and therefore it can be described by a set of forbidden graphs (minimal by relation "to be an induced subgraph" graphs not belonging to $\mathscr{X}$). Such characterization for P_2 is given by König theorem with known criterion for bipartite graphs. In addition to this classical theorem the following results are known for this type of simple graphs: In [3] it is described all forbidden subgraphs for the class of $\mathscr{K}(\mathscr{C})$, where $\mathscr{C}$ is the set of all simple cycles; In [1] it is found several families of forbidden graphs for $\mathscr{K}(P_3)$ and conjectured that the set of these families form a complete set of forbidden graphs for this class.

The aim of this work is to characterize the class $\mathscr{K}(P_4)$. There are two ways for giving such characterization. One of them is constructive: we show how to construct a graph of the given class by operations of edge subdivision and replacement of vertices with cographs. In the other approach we look for standard description of hereditary class by forbidden subgraphs. The set of forbidden subgraphs includes ten infinite families and 62 individual graphs. Whether it is a complete set of minimal forbidden subgraphs for (P_4) is an open question. We believe that it is complete.

In what follows covering and packing mean P_4-covering and P_4-packing, and König graph means König graph for P_4. The maximum number of subgraphs in P_4-packing of G is denoted as $pack(G)$, and the minimum number of vertices in its P_4-covering as $cover(G)$.

Induced subgraph P_4 is called quartet. We denote by (v_1, v_2, v_3, v_4) a quartet that consists of vertices v_1, v_2, v_3, v_4.

We denote by $|G|$ number of vertices in G.

Considering cycle C_n assume that the vertices are labeled along the cycle as $0, 1, \ldots, n-1$. The arithmetic operations with the vertex labels are performed modulo n. Each residue class of vertices numbers for modulo 4 is called 4-class.

2 Some Basic Properties

Definition 1 We say that a class of graphs is self-complementary, if for every graph it also contains its complementary graph.

Lemma 1 *The class of König graphs is self-complementary.*

Proof Since $P_4 \simeq \overline{P_4}$ each quartet in G corresponds to quartet in $\overline{G}$ induced by the same vertex set and vice versa. Therefore $pack(G) = pack(\overline{G})$ and $cover(G) = cover(\overline{G})$ for every graph G.

If G is König it means that it has packing and covering of equal cardinality and this property is inherited by all its induced subgraphs. But then graph $\overline{G}$ has the same properties too that is it is also König. □

Corollary 1 *If graph F is the minimal forbidden subgraph for class $\mathcal{K}(P_4)$, then $\overline{F}$ is also minimal forbidden graph for this class.*

Thus, it makes sense to look for a pairs of minimal forbidden subgraphs complementary one to another.

Definition 2 We say that a connected graph G is reduced if its complement is connected and each of its vertices belongs to at least one induced 4-path.

Since $\mathcal{K}(P_4)$ is self-complementary hereditary class, the graph G is König if and only if:

1. Each connected component of graphs G and $\overline{G}$ is a König graph;
2. The graph obtained from graph G by removing all vertices which don't belong to any quartet is König.

So to answer the question whether a graph is König it is enough to answer this question for each of its inclusion-maximal reduced subgraphs. We call them reduced components.

Below we consider only reduced graphs, but other types of graphs may appear in proofs.

3 4-Widening

Definition 3 A graph without quartets is called cograph.

Definition 4 The operation of replacement of vertex x with cograph consists of the following steps:

1. The vertex is removed from the graph.
2. Several new vertices are added to it. New vertices are interconnected so as to form a cograph.
3. Each of new vertices is connected by an edge to each vertex adjacent to x in the original graph.

Definition 5 We call an induced path of a graph terminal if one of its vertices is terminal (has degree 1), and the others have degree at most 2. We call vertex adjacent to a terminal path if it is adjacent to one of the vertices of the path but it does not belong to it (if exists).

Definition 6 The operation of replacement of a terminal path of n vertices with cograph ($n \leq 3$) consists of the following steps:

1. Vertices of this path are removed from the graph.
2. Several new vertices are added to graph. New vertices are interconnected so as to form cograph.
3. Let y be a vertex adjacent to path. New vertices are connected to the vertex y so that a maximum induced path which contains y and the added vertices have length n.

Definition 7 Firstly, we replace some vertices of degree 1 and 2 in G with cographs, then in the resulting graph we replace some terminal paths of at most 3 vertices with cographs. The resulting graph is called 4-widening of G, and the cographs of the replaced vertices are called sections. Each vertex which is not replaced is considered as a separate section. The cographs obtained by replacement of terminal paths are not sections. We call them terminal cographs.

Lemma 2 *Let A be a section in the graph G. A vertex $v \in A$ belongs to the minimum covering of graph G if and only if all the vertices of this section also belong to this minimum covering.*

Proof Assume that A contains vertex a which belongs to some of the minimum coverings and vertex b which does not belong to it. Since the covering is the minimum there exists a quartet in G, from which only vertex a belongs to the covering (otherwise this vertex could be removed from the covering). Then three other vertices of this quartet together with vertex b form an uncovered quartet, which contradicts the minimality of the covering. □

Thus, each minimum covering of any 4-widening of any graph consists of whole sections and possibly vertices of terminal cographs.

Suppose there exists the minimum covering of the graph which includes the vertex x, belonging to a terminal cograph. Since any quartet which includes the vertices of the terminal cograph includes the vertex y adjacent to the original path, replacing x with y, we also obtain the minimum covering. Thus, in any extension of any graph there exists the minimum covering consisting only of the whole sections. Below we consider only such coverings.

Lemma 3 *Any 4-widening of the tree belongs to $\mathcal{K}(P_4)$.*

Proof Let G be a 4-widening of a tree T. Cograph of G which replaced vertex x of T is denoted by $K(x)$.

Suppose there are two leaves adjacent to the same vertex in T. Then obviously the graph can be obtained from a tree by replacing a leaf with graph O_2. Thus, it suffices to consider only the trees in which each vertex is adjacent to at most one leaf.

Each terminal cograph can be considered as the set of sections corresponding to each of the vertices of the replaced path. Since we consider only coverings, consisting of the whole sections we can assume that the capacity of all fictitious sections is greater than 1.

We prove this by induction on the number of vertices in the graph G. The following cases exist:

Case 1. The number of vertices in T is less than 4. Then $pack(G) = cover(G) = 0$.

Case 2. The number of vertices in T equals 4. Then obviously $pack(G) = cover(G) = min(|K(x_1)|, |K(x_2)|, |K(x_3)|, |K(x_4)|)$.

Case 3. T has a quartet (x_1, x_2, x_3, x_4) such that $deg(x_1) = 1$, $deg(x_2) = 2$, $deg(x_3) > 2$, $deg(x_4) = 1$ or $deg(x_4) = 2$ and x_4 is connected to a leaf. Select vertices $a_1 \in K(x_1)$, $a_2 \in K(x_2)$, $a_4 \in K(x_4)$ and consider the graph G' obtained from G by removing the vertices a_1, a_2, x_3, a_4. Let P be the maximum packing and C be the minimum covering of the graph G'. By the induction hypothesis $|P| = |C|$. G' may be not reduced. In this case we can separate it to the reduced components and extend the induction hypothesis to each of them. Now $P \cup \{(a_1, a_2, x_3, a_4)\}$ is a packing and $C \cup \{x_3\}$ is a covering of the graph G.

Case 4. T has a quartet (x_1, x_2, x_3, x_4) such that $deg(x_1) = 1$, $deg(x_2) = deg(x_3) = 2$, $deg(x_4) \geq 2$. Select vertices $a_1 \in K(x_1)$, $a_2 \in K(x_2)$, $a_3 \in K(x_3)$, $a_4 \in K(x_4)$. Consider the graph G', obtained from G by removing these four vertices. Let P be the maximum packing and C be the minimum covering of the graph G'. By the induction hypothesis $|P| = |C|$. If we add quartet (a_1, a_2, a_3, a_4) to P, we get a packing of graph G with cardinality $|P| + 1$. We show that there exists a covering of the same cardinality in G. Non-empty sets $K(x_i) - \{a_i\}$, $i = 1, 2, 3, 4$, are the sections in the graph G'. If some set $K(x_i) - \{a_i\}$ is a subset of C (there can be only one such set), then $C \cup \{a_i\}$ is covering of the graph G. If none of these sets is the subset of C, then some of them are empty. Let i^* be the maximum i in which $|K(x_i)| = 1$. Then $C \cup \{a_{i^*}\}$ is covering of the graph G. □

Corollary 2 *Complement of any 4-widening of any tree belongs to the class $\mathcal{K}(P_4)$.*

4 Forbidden Subgraphs

As stated in Corollary 1 set of forbidden subgraphs for the class of König graphs can be split into pairs of complementary graphs.

We first consider several infinite series of forbidden subgraphs for $\mathcal{K}(P_4)$. Obviously,

$$pack(C_{4k}) = pack(C_{4k+1}) = pack(C_{4k+2}) = pack(C_{4k+3}) = k;$$

$$cover(C_{4k}) = cover(C_{4k-1}) = cover(C_{4k-2}) = cover(C_{4k-3}) = k$$

(the latter is valid for $k > 1$). Therefore, by Lemma 2 the following statement is true.

Proposition 1 *Cycle C_n belongs to $\mathcal{K}(P_4)$, if n is divisible by 4 and C_n is minimal forbidden graph for $\mathcal{K}(P_4)$, if n is not divisible by 4.*

Denote by $\mathscr{C}$ the set of all forbidden cycles and their complementary graphs.

Consider the graph obtained from the cycle C_n by adding two vertices which are not adjacent to each other each of which is connected by an edge with one vertex of the cycle. This graph is denoted by $A(n, k)$, where k is the distance between the vertices of degree 3.

Proposition 2 *$A(n, k)$ is a minimal forbidden graph for $\mathcal{K}(P_4)$ if and only if n is divisible by 4 and k is odd.*

Proof Let $n = 4t$. Obviously $pack(A(n, k)) = t$. Let M be the minimum covering of cycle C_n. Then M is a 4-class. Since k is odd one of the vertices in $A(n, k)$ adjacent to vertex of degree 1 is also adjacent to a vertex of M. But then the distance to the nearest vertex of M is 3. So this vertex, the vertex of degree 1 adjacent to it and two vertices of the cycle induce a quartet, which is not covered by vertices of M. Therefore, $cover(A(n, k)) > t$ and $A(n, k) \notin \mathcal{K}(P_4)$.

All induced subgraphs of $A(n, k)$ are forests except the graph consisting of a cycle with the added vertex of degree 1. But for this graph obviously $pack(G) = cover(G) = t$ and it is König too.

If k is even, then covering can be made up from 4-class of the original cycle that contains one of the vertices adjacent to a vertex of degree 1. Thus in this graph $pack(G) = cover(G) = t$ and it is not forbidden. □

Denote $\mathscr{A} = \left\{ A(n, k), \overline{A(n,k)} \mid n \text{ is divisible by 4, } k \text{ is odd} \right\}$.

We call $n - l$-lasso ($l \in \mathbb{N}, n \in \mathbb{N}, n \geq 3$) the graph obtained from an n-vertex cycle by adding terminal path of length l to the vertex of it. The unique vertex of degree 3 in this graph we call a knot.

Denote by $B_2(n, k)$ the set of graphs obtained from $n - 2$-lasso by replacing a cycle vertex at distance k from the knot with a two-vertex cograph (i.e. with K_2 or O_2).

Proposition 3 *Graphs of set $B_2(n, k)$ are minimal forbidden subgraphs for class $\mathcal{K}(P_4)$ if and only if n and k are divisible by 4 and $k \neq 0$.*

Proof Let $n = 4t$ and $B \in B_2(n, k)$. It is evident that $pack(B) = t$. It is easy to check that for $k = 4l$, $1 \leq l \leq \frac{t}{2}$ none of the four minimum coverings of cycle C_n is a covering for B, so $cover(B) > t$. All induced subgraphs of the graph except three subgraphs are 4-widenings of forests. One of them is $4t - 2$-lasso. For this graph it is obvious that $pack(G) = cover(G) = t$. It is therefore König. The other two consist of a cycle with the vertex replaced with cograph and an isolated vertex or a vertex adjacent to the knot of 4-widening of lasso. Consider the 4-class of a cycle containing vertices at distance 2 from a knot of a lasso. It is easy to see that such 4-class is a covering for both graphs so $pack(G) = cover(G) = t$, and they are also König. □

Denote by $B_1(n, k_1, k_2)$, where $k_1 \neq k_2$, the set of graphs obtained from the $n-1$-lasso by replacement of cycle vertices at distances k_1 and k_2 from the knot with cographs of two vertices (K_2 or O_2).

Proposition 4 *Graphs of set $B_1(n, k_1, k_2)$ are minimal forbidden subgraphs for class $\mathcal{K}(P_4)$ if and only if n and k_1 are divisible by 4, k_2 is even but not divisible by 4 and $k_1 \neq 0$.*

Proof Let $n = 4t$ and $B \in B_1(n, k_1, k_2)$. Obviously $pack(B) = t$. It is easy to check that for $k = 4l$, $1 \leq l \leq \frac{t}{2}$ none of the four minimum coverings of cycle C_n is a covering for B, so $cover(B) > t$.

All induced subgraphs of the graph are extended forests, except three. The first of them is a wided cycle. It is enough to take a 4-class of a cycle that contains no vertices replaced with cographs as the cover. So for this graph obviously $pack(G) = cover(G) = t$ and it is therefore König. The other two can be obtained from the n-1-lasso by replacing one vertex of the cycle with cograph. Consider the 4-class of cycle containing vertex at distance 2 from the vertex replaced with cograph. It is easy to see that such 4-class is covering for both graphs so $pack(G) = cover(G) = t$, and they are also König. □

Denote

$$\mathscr{B} = \bigcup_{\substack{n=4s, k=4t, \\ s,t \in \mathbb{N}}} \{G \mid G \in B_2(n,k) \text{ or } \overline{G} \in B_2(n,k)\} \cup$$

$$\cup \bigcup_{\substack{n=4s, k_1=4t, \\ k_2 = 4q-2, \\ s,t,q \in \mathbb{N}}} \{G \mid G \in B_1(n, k_1, k_2) \text{ or } \overline{G} \in B_1(n, k_1, k_2)\}$$

Denote by $D(k_1, k_2, k_3, k_4)$ the set of graphs obtained from a cycle of length $n = k_1 + k_2 + k_3 + k_4$ by replacement of four vertices numbered $0, k_1, k_1 + k_2, k_1 + k_2 + k_3$ with cographs of two vertices. This set always consists of exactly 16 graphs which differ by the structure of cograph, which replaced the vertex (K_2 or O_2).

Proposition 5 *Graphs of set $D(k_1, k_2, k_3, k_4)$ are minimum forbidden subgraphs for class $\mathcal{K}(P_4)$ if and only if $k_1 \equiv k_2 \equiv k_3 \equiv k_4 \equiv 1 \pmod 4$, $k_i \geq 5$, $i = 2, 3, 4$ or $k_1 \equiv 1 \pmod 4$, $k_1 \geq 5$, $k_2 \equiv k_4 \equiv 2 \pmod 4$, $k_3 \equiv 3 \pmod 4$.*

Proof Let $D \in D(k_1, k_2, k_3, k_4)$. Vertices numbered $0, k_1, k_1 + k_2, k_1 + k_2 + k_3$ in cycle are replaced with cographs. We denote the corresponding sections as A_0, A_1, A_2, A_3. Let $n = k_1 + k_2 + k_3 + k_4 = 4t$. Obviously $pack(D) \geq t$. We show that in fact the equality holds.

If the inequality is strict, then there exists a packing which separates all vertices into quartets. If $k_1 \geq 5$, then in both cases it necessarily includes quartet $(x, n-1, n-2, n-3)$ where $x \in A_0$ and (y, k_1+1, k_1+2, k_1+3), where $y \in A_1$. Thus the interval from 0 to k_1 has $k_1 + 1 \equiv 2 \pmod 4$ vertices and they all have to be separated into quartets, which is impossible.

If $k_1 = 1$ (here we consider only the first case), then sections A_0, A_1 are adjacent and the packing separating all the vertices into quartets must have quartet $(y, x, n-1, n-2)$ where $x \in A_0$, $y \in A_1$, and (z, k_2+2, k_2+3, k_2+4) where $z \in A_2$. Thus, the interval from 0 to $k_2 + 1$ includes $k_2 + 2 \equiv 3 \pmod 4$ vertices and they all have to be separated into quartets which is impossible.

There are 4-widenings of forests and cycle with one two or three vertices replaced with cographs among the induced subgraphs D. For each of these graphs it is obvious that $pack(G) = cover(G) = t$. □

Denote

$$\mathscr{D} = \bigcup_{\substack{k_1 \equiv k_2 \equiv k_3 \equiv \\ \equiv k_4 \equiv 1 \pmod 4, \\ k_i \geq 5,\ i = 2, 3, 4 \\ or \\ k_1 \equiv 1 \pmod 4,\ k_1 \geq 5, \\ k_2 \equiv k_4 \equiv 2 \pmod 4, \\ k_3 \equiv 3 \pmod 4}} \{G \mid G \in D(k_1, k_2, k_3, k_4)\ or\ \overline{G} \in D(k_1, k_2, k_3, k_4)\}$$

Consider the set of minimal forbidden graphs that are not part of any of the infinite families $\mathscr{A}, \mathscr{B}, \mathscr{C}, \mathscr{D}$. Direct verification can establish that among the graphs with not more than 7 vertices there are exactly 64 such graphs: $E_1, \ldots, E_{32}$ and their complements (Figs. 1 and 2). For each of them $pack(G) = 1$, $cover(G) = 2$. We denote $\mathscr{E} = \{E_1, \ldots, E_{32}, \overline{E_1}, \ldots, \overline{E_{32}}\}$.

The following theorem summarizes the results on the forbidden subgraphs of class $\mathscr{K}(P_4)$

Theorem 1 *The set of minimal forbidden graphs of class $\mathscr{K}(P_4)$ is infinite and contains the set $\mathscr{A} \cup \mathscr{B} \cup \mathscr{C} \cup \mathscr{D} \cup \mathscr{E}$.*

5 Widened Subdivisions of Bipartite Graph

Definition 8 Let H be a bipartite graph. Each cyclic edge (an edge which belongs to a cycle) of this graph is subdivided by one vertex. We replace each vertex added at subdivision and each vertex of degree 1 or 2 not belonging to the cycle with an arbitrary cograph. We also replace with a cograph some of the old vertices of degree 2 belonging to cycles except the following conditions:

König Graphs for 4-Paths

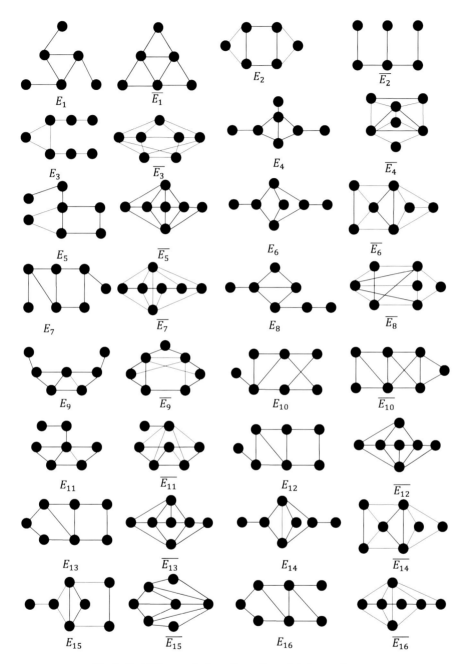

Fig. 1 Forbidden graphs from E_1 to E_{16} and their complements

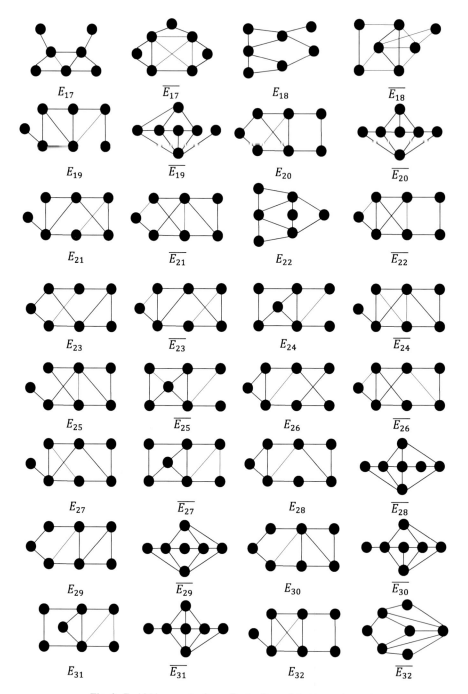

Fig. 2 Forbidden graphs from E_{17} to E_{32} and their complements

1. for each cycle if there is a vertex v adjacent to three or more vertices of degree more than 1 then any vertex of 4-class containing v cannot be replaced with cograph;
2. for each cycle if a there is a vertex v of degree 3 or more in the cycle then a vertex of the 4-class containing v and a vertex of another 4-class consisting of old vertices cannot be replaced with a cographs simultaneously.

We call the obtained graph a widened subdivision of the original bipartite graph.

Theorem 2 *Any reduced graph obtained by a widened subdivision from arbitrary bipartite graph which is not a simple cycle is König.*

Proof Let G be a reduced graph that is a widened subdivision of a bipartite graph H. If H is not a simple cycle, and it has cycles, then there is at least one vertex of degree 3 or more in each of these cycles. From the definition of a widened subdivision it follows that in every widened cycle of G there is at least one 4-class with no replaced vertices. Denote by A the set of all such vertices. For each of these vertices in G choose any quartet, consisting of this vertex and the three other vertices belonging to the corresponding cycle. Let B be the set of all selected quartets. If we delete from G all the vertices of A, then the remaining graph F is the 4-widening of forest. For F $pack(F) = cover(F)$ by Lemma 3.

Adding the set A to the minimum covering and the set B to the maximum packing of the graph F, we obviously obtain covering and packing of G having equal cardinalities. Thus, $pack(G) = cover(G)$. □

References

1. Alekseev, V.E., Mokeev, D.B.: König graphs with respect to 3-paths. Diskretnyi Analiz i Issledovanie Operatsiy **19**, 3–14 (2012)
2. Deming, R.W.: Independence numbers of graphs — an extension of the König-Egervary theorem. Discrete Math. **27**, 23–33 (1979)
3. Ding, G., Xu, Z., Zang, W.: Packing cycles in graphs II. J. Comb. Theory. Ser. B. **87**, 244–253 (2003)
4. Grötschel, M., Lovasz, L., Schrijver, A.: Geometric Algorithms and Combinatorial Optimization. Springer, Heidelberg (1993)
5. Hell, P.: Graph packing. Electron. Notes Discrete Math. **5**, 170–173 (2000)
6. Lovasz, L., Plummer, M.D.: Matching Theory. Akadémiai Kiadó, Budapest (1986)
7. Mishra, S., Raman, V., Saurabh, S., Sikdar, S., Subramanian, C.R.: The complexity of konig subgraph problems and above-guarantee vertex cover. Algorithmica **61**, 857–881 (2011)
8. Yuster, R.: Combinatorial and computational aspects of graph packing and graph decomposition. Comput. Sci. Rev. **1**, 12–26 (2007)

A Hybrid Metaheuristic for Routing on Multicast Networks

Carlos A.S. Oliveira and Panos M. Pardalos

Abstract Multicast routing systems have the objective of simultaneously transferring data to multiple destination nodes while using a single "push" operation. This leads to cost savings associated with reduced bandwidth utilization, which results from a decrease in data duplication across network links. An important problem on multicast networks, known as the delay constrained multicast routing problem (DCMRP), asks for the determination of an optimal route for packet transfers between members of a multicast group. Several heuristics have been proposed in the last few years to solve the DCMRP, which is of great interest for telecommunication engineers. In this paper we propose a novel, hybrid metaheuristic approach for the DCMRP, where a greedy randomized adaptive search procedure is used along with variable neighborhood search algorithm to find near optimal solutions. Computational experiments show that the proposed technique provides superior solution quality, while it is also efficient in terms of the use of computational resources.

1 Introduction

Multicast services have been used in modern network applications to allow direct communication between a source node and a set of receivers, referred to as multicast destinations [2, 13]. In recent years, the number of applications of multicasting has increased steadily, following the rapid advances in the availability and use of the Internet as well as intranets in the corporate world. Multicast networks are known

---

C.A.S. Oliveira (✉)
Quantitative Research Dept., F-Squared Inc., Princeton, NJ, USA
e-mail: oliveira@ufl.edu

P.M. Pardalos
Department of Industrial and Systems Engineering, Center for Applied Optimization, University of Florida, Gainesville, FL, 32608, USA

Laboratory of Algorithms and Technologies for Networks Analysis, National Research University Higher School of Economics, Nizhny Novgorod, 603155, Russia
e mail: pardalos@ufl.edu

to provide robust and efficient data delivery for a wide spectrum of applications, including video-on-demand, groupware, and data streaming, among others [14, 22].

A number of algorithmic issues, however, remain as a major problem for the wide adoption of multicasting networks. For example, routing is an issue that has not been completely resolved on such systems due to the high computational cost of exact algorithms. While for traditional unicast systems the routing problem can be solved in polynomial time using well-known methods such as the Dijkstra's algorithm [3], multicast routing is better modeled by the Steiner tree problem, which is one of the basic NP-hard problems [7, 17].

Given the inherent complexity of exact approaches to the DCMRP, a large number of heuristics have been proposed to find good, if non-optimal, solutions that can be calculated in polynomial time [2, 6, 10, 11, 14]. However, many of these local search methods proposed in the telecommunications engineering literature suffer from a lack of optimality guarantees and may be easily trapped into local optima [19–21, 23]. As such, these methods are indicated only for application to small- to medium-scale instances. On the other hand, applications of the DCMRP become even more challenging for instances with a large number of multicast members, since the efficient use of resources turns into a critical factor for the success of such network implementations.

In this paper we propose a metaheuristic solution for the delay constrained multicast routing problem. In particular, we propose a new method for computing routing trees for multicast networks using a hybridization of greedy randomized adaptive search procedure (GRASP) and variable neighborhood search (VNS). The strategy here is to improve the performance of the algorithm by avoiding spending too much time exploring suboptimal solutions and their solution spaces.

At the same time, our contribution may be extended to the general application of a hybrid GRASP metaheuristic. By combining the general structure of GRASP with VNS, the result is a novel search algorithm that may be used to produce fast implementations for several related problems.

The paper is organized as follows. In the next section (Sect. 2), we provide a detailed definition for the DCMRP, using graph theoretical concepts as well as a mathematical programming model. Then, in Sect. 3 we develop a GRASP metaheuristic for the DCMRP, followed by a description of the VNS strategy employed. We present computational results for our approach in Sect. 4, and finally some concluding remarks are provided in Sect. 5.

2 Delay-Constrained Multicast Routing

Multicast networks have been designed with the explicit goal of allowing fast data transmission from a source node to a set of destinations, while using a single send operation. This is made possible by sending data only once over a network link whenever one or more destinations have requested the same content. A set of nodes interested in a particular piece of data is called a *multicast group*. The main task faced in the operation of a multicast network consists of delivering the requested data to all members of a multicast group. To accomplish this goal, the system needs to determine a set of routes connecting sources to destinations.

Let $G = (V, E)$ be a graph where V is the set of nodes and E a set of links connecting adjacent nodes. The source is denoted by s with destinations $D = \{d_1, \ldots, d_k\}$, such that $D \subset V$. The cost function $c : E \rightarrow Z_+$ represents link costs and the delay function $\tau : E \rightarrow Z_+$ returns the time $\tau(e)$ elapsed when traversing edge $e \in E$. We also denote by $c(E')$ the cost associated with a set of edges $E' \subseteq E$, that is, $c(E') = \sum_{e \in E'} c(e)$. Similarly, we denote the time delay for path $\mathscr{P}$ in G as $\tau(\mathscr{P}) = \sum_{e \in \mathscr{P}} \tau(e)$.

The DCMRP asks for a set of edges $E' \subseteq E$ such that s is connected to every node $d \in D$ on $G' = (V, E')$, and the maximum acceptable delay Δ_d at destination d is a constant, i.e.,

$$\sum_{e \in \mathscr{P}_d} \tau(e) \leq \Delta_d, \quad \text{for every destination } d \in D,$$

where $\mathscr{P}_d$ is the path induced by E' in G, connecting s to d. Moreover, we require that the total cost $\sum_{e \in E'} c(e)$ of the subset E' be minimum.

Additionally, real-world instances of the DCMRC problem frequently have extra requirements on the kind of paths that can be used to connect sources and destinations [2]. For example, the model may require that a minimum capacity be available for each edge selected in the final solution. We will consider variations of this problem in the next sections. First, let us define a formal MIP model for the problem.

2.1 MIP Model for the DCMRP

A mixed integer programming (MIP) model for the DCMRP can be described as follows. Let $x_i \in \{0, 1\}$, for $i \in \{1 \ldots |E|\}$, be a decision variable that is 1 whenever an edge is part of the routing tree and 0 otherwise. Then, the objective function can be written as a minimization problem over the vector x, while considering the cost of each edge:

$$\min \sum_{e_i \in E} x_i c(e_i),$$

where $c : E \rightarrow \mathbb{R}$ is the cost function as described above. Then, we need a set of constraints that guarantee the connectedness of the solution set $\{(u, v) \in E : x_j = 1\}$:

$$\sum_{e_i = (v,w) \in U \times V} x_i \geq 1 \quad \forall \text{ partitions } U, W \text{ s. t. } |(S \cup D) \cap U| \text{ is odd.}$$

That is, there is at least one link connecting each partition of V where the number of sources and destinations is different. Next, we have constraints that indicate the boundedness of the delay.

$$\sum_{e_i \in E} y_i^v \tau(e_i) \leq \Delta_v \quad \text{for } v \in D$$

where y_i^v for $e_i \in E$ is an indicator variable with value 1 whenever the edge e_i is part of a path from source to destination $v \in D$. The variables y_i can be 1 only if link e_i is part of the solution, so we also have

$$y_i \leq x_i \quad \text{for all } e_i \in E.$$

Finally, we need to apply the standard integrality constraints to our model variables:

$$y_i \in \{0, 1\} \quad \text{for } e_i \in E$$

$$x_i \in \{0, 1\} \quad \text{for } e_i \in E$$

2.2 DCMRP and the Steiner Problem

The minimum routing cost problem as described above has close connections to the minimum cost Steiner tree problem. In graph theory, a tree that connects a set of required nodes, while using other nodes only if necessary, is called a Steiner tree [8, 16]. Thus, we can restate the problem as that of finding a minimum cost Steiner tree such that maximum delay restrictions are also satisfied.

In the Steiner problem, one is given a graph $G = (V, E)$ together with a cost function $c : E \to Z_+$, and a set $R \subset V$ of required nodes. The nodes in $V \setminus R$ are called Steiner nodes. The objective is to find a tree T linking the nodes in D, passing through Steiner nodes ($V \setminus R$) if necessary, such that the cost $\sum_{e \in T} c(e)$ is minimized. The Steiner problem on graphs is well known to be NP-hard [7].

Consider the following transformation from instances of the Steiner problem to the DCMRP. Given an instance of the minimum cost Steiner tree problem, let us construct an instance of DCMRP using the same underlying graph. Select a node among the required nodes to become the source and let the remaining required nodes be destinations. Then, set $\Delta_d \leftarrow \infty$, for all $d \in D$. As can be easily confirmed, an optimal solution to the transformed problem will also be a solution to the original instance of the Steiner problem. Conversely, an optimal solution for the original instance will give an optimal solution for the transformed instance. This argument shows that the DCMRP is also NP-hard.

Given the computational complexity of the DCMRP, it is extremely difficult to solve general large-scale instances of the problem. However, our goal is to devise algorithms that can provide near optimal solutions for typical instances of

```
Read instance data
Initialize GRASP data structures
S* ← ∅
while termination criterion not satisfied do
    S ← new greedy randomized solution
    S ← LocalSearch(s)
    if cost(S) < cost(S*) then
    |   S* ← S
    end
end
return S*
```

Algorithm 1: Generic GRASP algorithm

the problem. Moreover, we want such algorithms to perform efficiently, returning feasible solutions quickly and avoiding getting stuck in local optima.

3 GRASP Approach for DCMRP

Most modern metaheuristic techniques are based on finding solutions close to the global optimum with the help of gradient methods combined with randomization rules, which are designed to avoid local optima. Metaheuristics main contribution is on the development of intelligent strategies for mixing existing non-optimal techniques and algorithms. Such metaheuristics use similar principles in slightly different ways. A conceivable goal for the algorithm designer is, therefore, to borrow techniques from different metaheuristics, in order to create algorithms that better reflect the characteristics of the problem at hand. In this paper we use some of the techniques proposed by GRASP metaheuristic as well as by the variable neighborhood search metaheuristic (VNS) to efficiently solve the DCMRP.

Greedy Randomized Adaptive Search Algorithm (GRASP), proposed by Feo and Resende [4], aims at finding near optimal solutions for combinatorial optimization problems. It is composed of a number of iterations, where a new solution is picked from the available feasible set, using a greedy construction algorithm. The initial solution is subsequently improved using some local search method. GRASP has been very successful in a number of applications such as QAP [15], Frequency Assignment [9], Satisfiability, and many others [5]. The steps of standard GRASP are summarized in Algorithm 1.

The GRASP algorithm is a multi-start method, where a new solution is constructed, and subsequently improved. In the construction phase, at each iteration the algorithm tries to add a new element using a randomized greedy strategy. The second phase is concerned with improving the current solution. Its goal is to achieve a local optimum state by performing a local search, which is usually based on a gradient decent strategy. In the next sections, we describe the approach used in the GRASP implementation for DCMRP.

```
S ← ∅
while solution s is not feasible do
    Sort e₁, ..., eₖ using greedy function
    Select a random α, such that 0 < α ≤ k
    R ← {e₁, ..., eα}
    e ← arbitrary selected element of R
    S ← S ∪ {e}
end
return S*
```
Algorithm 2: Generic GRASP constructor

3.1 GRASP Constructor

The construction phase of the GRASP algorithm is responsible for creating a new solution, using a greedy randomized strategy. The basic idea behind greedy randomization is to add elements to the solution according to a greed criterion. However, the element that is chosen at each step to compose the new solution is not necessarily the best available element. Instead, the selection is taken randomly from a subset of best available elements, which have been previously sorted using a greedy objective. The general algorithm is displayed in Fig. 2. In the algorithm, the subset of best elements is called a restricted candidate list (RCL). The RCL is created and used to track the best elements available to be added to the current solution.

3.2 Speeding Up the GRASP Constructor

To be effective as the construction phase in a GRASP algorithm, it is desirable that the construction method be very fast. A construction algorithm that is not quick enough may become a bottleneck for the whole algorithm, since it needs to be executed every time a new solution is desired. While the traditional approach for GRASP construction works well, it requires the maintenance of an additional data structure, which needs time to build, and as a result it can waste computational resources. Instead, we use the following observation proved in [14].

Observation 1 *Let $x_1, \ldots, x_n$ be an unordered sequence, and $y_1, \ldots, y_n$ the corresponding ordered sequence. Then, to find a random element among the $y_1, \ldots, y_\alpha$, for $0 < \alpha \leq n$ is on average equivalent to selecting the best of α random elements of $x_1, \ldots, x_n$.*

The observation above gives a very efficient way of implementing the RCL test, which gives us, on average, the same results. Start with the full set C of candidate elements. Then, at each step generate a value of α, and pick at random $k = 1/\alpha$

```
Input: parameter α, instance size N
S ← ∅
while S is an incomplete solution do
    α' ← uniform(0, α)
    k ← N/α;
    c ← ∞ (−∞ for maximization problems)
    for j ∈ {1, ..., k} do
        C_j ← set of candidates at this iteration
        x ← arbitrary element from C_j
        if c(x) < c then
            c ← c(x)
            y ← x
        end
    end
    S ← S ∪ {y}
end
return S
```
Algorithm 3: Improved construction phase for GRASP

elements of C. From the picked elements, store only the one that is the best fit for the greedy function. This method is shown in Algorithm 3.

A clear advantage in terms of computational complexity is achieved by the proposed construction method for GRASP. The greatest advantage is that, while in the original technique the candidate elements must be sorted, this is not necessary in the proposed algorithm. Moreover, the complexity of traditional construction is dependent on the number of candidate elements. In our method, the complexity is constant for a fixed value of α. For example, if alpha is $n/2$, then we need just two iterations to find an element in the RCL, with high probability.

Theorem 1 ([14]). *The complexity of selecting elements from the RCL in the modified construction algorithm is $n \log n$.*

3.3 GRASP Construction Phase for DCMRC

We proceed to describe how to find a solution for the DCMRP that will be used in the construction phase of the GRASP algorithm. The construction algorithm is composed of several steps, in which we build a spanning tree that contains all the nodes in the required set of sources and destinations, along with other nodes required as intermediaries.

The first part of the construction phase is to compute paths connecting the source to each of the destinations $d \in D$. This is done using a randomized version of Dijkstra's algorithm, which finds shortest paths from a source to a single destination.

```
S ← ∅
while there is d ∈ D that is unconnected in S do
    𝒫 ← randomized shortest path from s to d
    S ← S ∪ 𝒫
end
while the is at least one cycle in s do
    e' ← arg max_{e∈s} c(e) s.t. e is contained in a cycle
    e'' ← arg max_{e∈s\e'} c(e) s.t. e is contained in a cycle
    ê ← e' with prob. p, otherwise e''
    S ← S \ ê
end
return S
```

Algorithm 4: GRASP construction for DCMRP

The adaptation necessary here is such that the shortest path is updated only with high probability p. This guarantees that the solution found in any two executions of the algorithm will be close to the optimum, but still with random differences that make the result useful for our stochastic optimization methods. The randomized shortest path is then used to create an initial solution as shown in Algorithm 4.

The first part of Algorithm 4 guarantees that the solution created is connected, by finding a separate path from the source to each destination $d \in D$. At the end of this phase, we will have a solution where separate paths may result in cycles. Therefore, the second phase of the construction algorithm aims at removing such cycles. At each step, it finds the two highest cost edges contained in a cycle. Then, the algorithm removes one of them in an arbitrary way. The goal is to reduce the cost of the final solution, while at the same time allowing for randomized results. The resulting solution is returned at the end of Algorithm 4 to be later used by GRASP.

3.4 Variable Neighborhood Search

Once a feasible solution has been created by GRASP, the next step of our algorithm is to try to improve its quality using a local search procedure. Traditionally, local search has been performed using gradient descent techniques, which try to incrementally improve a solution until a local minimum is reached. The disadvantage of such methods, however, is that they can quickly become stuck in a local neighborhood, hindering the computational effort employed during the search phase. We try to avoid this behavior by using instead an alternative search technique based on variable neighborhood search (VNS).

VNS is a metaheuristic programming technique that has be successfully used to solve several combinatorial optimization problems [12]. Its main approach is to perform local search using successively larger neighborhoods, until no more improvements can be found by increasing the neighborhood size up to a given parameter. The advantage of VNS is that it will not stop once the first local optimum

A Hybrid Metaheuristic for Routing on Multicast Networks

```
Input: current solution S
S* ← s /* initialize best solution */
N ← 1 /* set distance to 1 */
while N < K do
    N ← N + 1
    while improvement found in last δ iterations do
        u, v ← random pair or nodes in S such that $d_S(u, v) = N$
        $\mathscr{P}$ ← randomized shortest path u → v in G
        S' ← S with $\mathscr{P}$ in place of path u → v
        if c(S') < c(S*) then
        |   S* ← S'
        end
    end
end
return S*
```
Algorithm 5: Variable neighborhood search

has been found. Instead, it will continue to build better solutions by reaching more distant neighborhoods in a organized fashion, until it cannot find any additional improvements. By reaching these more distant neighborhoods, VNS can more easily avoid being trapped in the same optimal solution, therefore yielding better results than standard local search.

The VNS algorithm for the DCMRP tries to replace existing sub-paths in the current solution by using alternate paths that have the potential for improving the objective function. This is done by arbitrarily selecting a pair of nodes occurring in the existing solution and replacing its induced sub-path by a new path, found using a shortest path algorithm. For this purpose, one can use a method such as Dijkstra's shortest-path procedure, which will hopefully provide a local improvement to the existing solution.

It is also possible to use randomized shortest-paths, similarly to how these paths are calculated in the GRASP constructor, where the best path is updated according to a probability distribution, so that the best path is not always selected. In that case, possible improvements will come from the exploration of the neighborhood of the existing solution, although the local-optimality of this improvement is not guaranteed in the same form as when using a completely greedy procedure.

The difference between VNS compared to other local search strategies resides in the ability to change the underlying neighborhood structure as a new local minimum is found. In our case, the neighborhood is changed by increasing a reach parameter, so that larger subpaths are substituted by the algorithm. This kind of change will happen until the maximum of N, defined by the parameter K, is reached. The general approach used in our VNS implementation is provided in Algorithm 5.

The algorithm starts by defining the parameter N, which is interpreted as the distance between nodes for which a new path will be searched. For the first iteration, this parameter is set to value 1, and in this case only nodes that are neighbors in the current solution are substituted by new paths. This process is repeated as long as we are able to perform improvements in the existing solution. A limit of δ moves without improvement is allowed before the inner part of the algorithm is interrupted. Then, the parameter N is increased by one and the process restarts.

The search is completed only when N reaches the previously defined upper limit K. At this point, the algorithm has systemically investigated all paths that might improve the current solution by replacement of paths occurring between pairs of nodes. When that happens, we return to the upper-level GRASP process the best solution found during all iterations, which will be used as the new optimum solution.

3.5 Path Relinking

GRASP has the advantage of being easy to develop, since it is composed of relatively independent procedures (the constructor and local search phases). It is well suited for applications with existing heuristic algorithms, that can be combined with GRASP to find a better solution.

However, one of the weaknesses of GRASP is its incapacity to integrate good solutions found previously into the current search iteration. Since each iteration will create a completely different solution, there is no information added to the system when a good solution is found.

A method that has been used lately to overcome this problem is called *path relinking* (PR) [1,18]. In PR, a subset of the best solutions found is kept in a separate memory, called the *elite set*. At each iteration, one of the solutions s will be selected, and a process of comparing the current solution with s will start. Each component of the solution will be changed to the corresponding value on s, and after this a local search will be initiated to check for local optimality.

The disadvantage of PR is the large time it takes to run, in comparison with the rest of the GRASP algorithm. This, in practice, has been a restricting factor in the use of PR on practical applications. Although PR brings a relative boost in solution quality on each iteration, the effect can be negative since the number of iterations may be reduced due to the added complexity. Therefore, we need to use experimental data to determine the best trade-off between computational time and quality of results produced by path relinking.

The path relinking implementation used to solve the DCMRP is described in Algorithm 6. The first step of path relinking is to create a set of elite elements, denoted by $\mathscr{E}$. The maximum size of this set is given by ϵ, therefore for the first ϵ iterations we simply run the existing GRASP algorithm and add the resulting solution to the elite set.

When the elite set is complete, we start using it to perform improvements to the solution generated by GRASP. This is depicted in the second `while` loop in Algorithm 6. The termination criterion is usually based on number of iterations, time, or the distance to a known lower bound. The first step of the loop is to create a new solution using GRASP. Then, an element of $\mathscr{E}$ is arbitrarily selected and subsequently used during the path relinking process. For each path $\mathscr{P} \in s$ going

```
ℰ ← ∅
while |ℰ| < ϵ do
    S ← GRASP
    ℰ ← ℰ ∪ {S}
end
while termination criteria not satisfied do
    S ← GRASP
    S' ← arbitrary element of ℰ
    for each path 𝒫 ∈ S' do
        S ← S ∪ 𝒫
        while s has a cycle do
            e ← arg max_e c(e) such that S \ e is connected
            S ← S \ e
        end
    end
    /* Update Elite Set ℰ */
    γ* ← arg max_{γ∈ℰ} c(γ)
    if c(s) < c(γ*) then
        ℰ ← (ℰ ∪ {s}) \ γ*
    end
end
return γ*
```

Algorithm 6: Path relinking improvement phase

from source to destination, the algorithm will try to replace the edges of the existing solutions with the edges of the same path in the elite solution. This is performed in the following way: first, we calculate the union of the two edge sets. Then, we proceed to remove redundant edges in a greedy fashion. That is, for each high cost edge, we check if we can remove it while maintaining a connected solution. This process continues until there are no cycles in the resulting solution.

The last step of the algorithm loop is to update the elite set. For this purpose, we retrieve the worst solution $γ$ in the elite pool. If the current solution s improves on the cost of $γ$, then we replace $γ$ with s in the elite set. These steps are then performed until a predefined termination criterion is satisfied.

4 Computational Results

To test the quality of the proposed metaheuristic, we designed a set of DCMRP instances. The instances range in size from 40 to 100 nodes, which is representative of medium-size problems occurring in large companies or in clusters of medium-sized organizations. Edges have been added to these test networks with costs that are uniformly distributed between 1 and 10. The distances are assumed to be Euclidian, but can be easily adapted to other metrics such as Manhattan distances (Fig. 1).

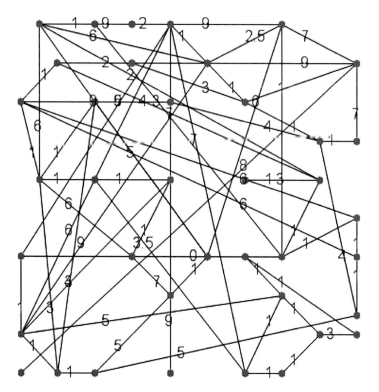

Fig. 1 Drawing of an instance with 40 nodes depicting node positions and costs between nodes

We run the proposed algorithm 10 times for each of the tested instances. We report on the average of these executions, to account for random fluctuations between different runs. The results are illustrated in Table 1. In this table, the first two columns give the number of nodes and edges in the network. The next three columns display the average best solution found by the GRASP without enhancements, the GRASP plus VNS strategy, and finally the GRASP with VNS enhanced with the PR method.

An area that we tested in the GRASP implementation just shown was the relative contribution of having two improvement methods (traditional local search and VNS) as components of the metaheuristic. As can be seen from the results, VNS is able to improve the quality of results in most of the cases, which shows that the use of multiple neighborhoods can provide a boost in efficiency for the algorithm.

Table 1 Experimental results of the GRASP and VNS metaheuristic implementation for the DCMRC

n	m	Best GRASP	Best GRASP+VNS	Best GRASP+VNS+PR	Time (s)
40	68	83	82	82	4.2
45	79	161	161	161	4.9
50	97	238	235	235	5.9
55	126	262	257	255	5.7
60	153	385	376	373	7.2
65	189	823	814	814	7.5
70	213	647	644	641	8.9
75	252	743	741	740	9.3
80	346	722	709	709	10.4
85	402	827	815	812	10.7
90	522	823	808	804	12.5
95	693	1028	1016	1009	13.3
100	816	1317	1311	1279	18.2

5 Concluding Remarks

In this paper, we presented a metaheuristic approach to solve the DCMRP, a problem arising on multicast routing systems, where the goal is to provide quick and accurate routing services to a set of source and destination points. Due to its occurrence in the design of telecommunication networks, the DCMRP has become the focus of intense research in the last few years. Our main contribution is the use of a fast construction algorithm along with an improvement method based on variable neighborhood search. The VNS has been used to enhance GRASP results, making it possible to explore existing solutions even faster.

The results of our experiments show that this method provides high quality solutions for realistic instances of the problem. The elegance of the method used also means that it can be easily incorporated to other algorithms for the DCMRP and related problems. In future research, it would be interesting to investigate the use of other intensification strategies for the proposed algorithm, such as improving the basic path relinking scheme used in this paper.

References

1. Aiex, R.M., Binato, S., Resende, M.G.C.: Parallel GRASP with path-relinking for job shop scheduling. Parallel Comput. **29**, 393–430 (2003)
2. Ballardie, A., Francis, P., Crowcroft, J.: Core-based trees (CBT) – an architecture for scalable inter-domain multicast routing. Comput. Comm. Rev. **23**(4), 85–95 (1993)

3. Dijkstra, E.W.: A note on two problems in connexion with graphs. Numerische Mathematik **1**(1), 269–271 (1959)
4. Feo, T.A., Resende, M.G.C.: Greedy randomized adaptive search procedures. J. Global Optim. **6**, 109–133 (1995)
5. Festa, P., Resende, M.G.C.: An annotated bibliography of grasp, part ii: Applications. Int. Trans. Oper. Res. **16**, 131–172 (2009)
6. Ganjam, A., Zhang, H.: Internet multicast video delivery. Proc. IEEE **93**(1), 159–170 (2005)
7. Garey, M.R., Johnson, D.S.: Computers and Intractability: A Guide to the Theory of NP-completeness. W.H. Freeman and Company, San Francisco (1979)
8. Gilbert, E.N., Pollak, H.O.: Steiner minimal trees. SIAM J. Appl. Math. **16**, 1–29 (1968)
9. Gomes, F.C., Pardalos, P.M., Oliveira, C.A.S., Resende, M.G.C.: Reactive GRASP with path relinking for channel assignment in mobile phone networks. In: Proceedings of the 5th International Workshop on Discrete Algorithms and Methods for Mobile Computing and Communications, pp. 60–67. ACM Press, New York (2001)
10. Hong, S., Lee, H., Park, B.H.: An efficient multicast routing algorithm for delay-sensitive applications with dynamic membership. In: Proceedings of IEEE INFOCOM'98, pp. 1433–1440 (1998)
11. Kompella, V.P., Pasquale, J.C., Polyzos, G.C.: Optimal multicast routing with quality of service constraints. J. Network Syst. Manag. **4**(2), 107–131 (1996)
12. Mladenović, N., Hansen, P.: Variable neighborhood search. Comput. Oper. Res. **24**(11), 1097–1100 (1997)
13. Oliveira, C.A.S., Pardalos, P.M.: A survey of combinatorial optimization problems in multicast routing. Comput. Oper. Res. **32**(8), 1953–1981 (2005)
14. Oliveira, C.A.S., Pardalos, P.M.: Mathematical Aspects of Network Routing Optimization. Springer, New York (2010)
15. Oliveira, C.A.S., Pardalos, P.M., Resende, M.G.C.: GRASP with path-relinking for the QAP. In: 5th Metaheuristics International Conference, pp. 57.1–57.6, Kyoto, Japan, August (2003)
16. Pardalos, P.M., Du, D.-Z., Lu, B., Ngo, H.: Steiner tree problems. In: Floudas, C.A., Pardalos, P.M. (eds.), Encyclopedia of Optimization, vol. 5, pp. 277–290. Kluwer Academic, Dordrecht (2001)
17. Pardalos, P.M., Khoury, B.: A heuristic for the steiner problem on graphs. Comput. Optim. Appl. **6**, 5–14 (1996)
18. Resende, M.G.C., Ribeiro, C.C.: A GRASP with path-relinking for private virtual circuit routing. Networks **41**, 104–114 (2003)
19. Salama, H., Reeves, D., Viniotis, Y.: A distributed algorithm for delay-constrained unicast routing. In: Proc. IEEE INFOCOM'97, Kobe, Japan (1997)
20. Sriram, R., Manimaran, G., Siva Ram Murthy, C.: A rearrangeable algorithm for the construction of delay-constrained dynamic multicast trees. IEEE/ACM Trans. Network. **7**(4), 514–529 (1999)
21. Yang, D.-N., Liao, W.J.: On bandwidth-efficient overlay multicast. IEEE Trans. Parallel Distr. Syst. **18**(11), 1503–1515 (2007)
22. Yang, Y., Wang, J., Yang, M.: A service-centric multicast architecture and routing protocol. IEEE Trans. Parallel Distr. Syst. **19**(1), 35–51 (2008)
23. Zhu, Q., Parsa, M., Garcia-Luna-Aceves, J.J.: A source-based algorithm for delay-constrained minimum-cost multicasting. In: Proc. IEEE INFOCOM95, pp. 377–385 (1995)

Possible Ways of Applying Citations Network Analysis to a Scientific Writing Assistant

Alexander Porshnev and Maxim Kazakov

Abstract Development of linguistic technologies gave rise to a new type of tools for academic writing, which use natural language processing and heuristics to help authors write scientific papers. In our contribution we present a new function "advise a paper to read" and the way it could be implemented. We discuss a possibility of using different centrality metrics and test their application in 50 cases created from 50 top cited articles of the Engineering domain. For each case we created a citation network graph based on the results of a search query in the Web of Knowledge by Thomson Reuters, using adjusted authors key phrases, and compared the results of applying the centrality metrics with the actual reference list presented in each article.

1 Introduction

Development of the language science and linguistic technologies led to creating powerful search engines and artificial intelligence systems to assist humans in many tasks. For example, Wolfram Alpha, one of the first computational and knowledge engines, allows the user to ask questions in natural language. This also led to developing automatic scientific writing assistants, a tool that helps writers revise and enhance their articles.

We should mention "A First-Language-Oriented Writing Assistant System" (FLOW) developed by Chen and coauthors, which allows the user to write by providing suggestions for paraphrases, automatically translating words into English or presenting more frequent n-grams [5].

Wu et al. created a "Collocation Inspector" for automatic collocation suggestion in academic writing, which helps non-native English writers tackle word usage problems [5]. The "Collocation Inspector" provides automated suggestions for verb–noun lexical collocations, based on contextual information analyzed by

A. Porshnev (✉) • M. Kazakov
National Research University Higher School of Economics, Nizhny Novgorod,
Russian Federation
e-mail: aporshnev@hse.ru; max.a.kazakov@gmail.com

machine learning methods. Chen et al. write about PREFER, a tool for Generate Paraphrases, based on the graphs approach [5]. Although those tools could support academic writing, they focused on the proper use of the English language and did not provide the author with advice on how to improve his paper. The only solution capable of analyzing a scientific text by using natural language processing and powerful heuristics is SWAN—Scientific Writing Assistant. In SWAN the ideas and experience of Lebrun were implemented by a team of Finnish computer scientists [11]. SWAN allows the user to analyze Title, Keywords, Introduction, and Conclusion sections and provides tips on how a paper could be improved. The main aim of SWAN is to help the user enhance the readability of a paper. Although SWAN assesses the fluidity and cohesion of a contribution, it does not provide ideas about how the user could improve scientific research by looking from a wider perspective, which is what a scientific supervisor can usually do.

In this paper we focus on suggesting to a writer what else he probably should read to improve the quality of his paper. This task is usually encountered by the writer in a new research area. A scientific supervisor can normally provide an overview for a field, but sometimes, when starting research in a new interdisciplinary field, even a scientific supervisor can face problems. How can a new area be analyzed to find the most competent authors and valuable contributions? To handle this task, a couple of good reviews normally need to be found, but how can we do that with no reviews available? Another way to enter a new scientific area is to browse through citation lists to discover the most relevant papers by title, or to look at the highest cited papers. It may be very successful, but maybe other metrics could provide even better results. Following the ideas of Ma, Guan, Zhao, we decided to analyze how analysis of the citation network can help find the most competent authors and valuable contributions [13].

1.1 Related Work

Citation analysis may be classified to pertain to two main areas. The first one involves studying bibliometrics, research trends, and impacts of scientists (CiteSpace, BiblioTools2, Sci2, etc.). The second one is connected with scientific search engines, for example CiteSeerX, Google Scholar, etc., accumulates a huge amount of articles and allows better search of scientific literature. The middle area between these two approaches is where efforts are undertaken to extend analysis techniques to improve search queries or provide hints for researchers.

Works in bibliometrics give a detailed description of networks, using different graph-based methods (e.g., [10]), mapping and labeling research areas (e.g., [4]). Some of them offer ideas how certain centrality metrics could be used to improve research quality. For example, while analyzing citation networks, Chen suggested that betweenness centrality could be used to find out the potential pivotal points of paradigm shift [4].

There are several projects devoted to citation network analysis for bibliometrics (for example, CiteSpace, Sci2, etc.). CiteSpace was created by Chen to visualize the citation networks. Although focusing mainly on detecting and visualizing trends and transient patterns in scientific literature, CiteSpace provides impressive visualization possibilities and can be employed to create and visualize cited references networks [4]. CiteSpace, however, is unable to create a directed cited reference graph, in which the source node will be an article and target nodes will be papers cited in this article.

The area of scientific search engines is less presented in scientific literature, and most of the used algorithms are not explained. For example, we could not find Google Scholar's ranking algorithm. Nor could we find any research on evaluation of search query quality for scientific databases.

In the middle area of further development of search techniques and recommendations, which is the primary point of our interest, we found a work by Ma, Guan, and Zhao, who discussed possible use of PageRank in Citation Analysis [13]. They suggested that PageRank could provide a more integrated picture of publications' influence in a specific area [13]. In 2000–2005, while analyzing the articles published in 261 magazines of the "Biochemistry and Molecular Biology" domain, they supposed that the PageRank-based approach could help identify several influential papers suffering from low citations, and further make them visible to the research community [13]. Thus they found the results obtained with PageRank to be highly correlated with citation counts (0.9), which means PageRank can be considered a reliable indicator representing the importance of scientific publications [13]. Although their analysis only revealed one disparity between PageRanks and Internal Citations, we wonder if we will discover the same tendency in more cases.

2 Methodology

Analysis of citation networks, scientometrics, and bibliometrics normally uses a co-occurrence graph, but the resources like Web of Science or Science Direct can also provide information about a citation list (cited references). Using the citation list we can create a citation graph which is directed and has papers as its nodes. Edges represent a citation relationship, and if an author cites paper B in paper A, we will have a directed edge from A to B.

Our point of view is that the citation graph and analysis of its structure may provide more information than a simple request for the most cited paper, which raises the following question: What is the most useful graph centrality metrics?

To find the most useful centrality metrics, we analyzed highly cited papers of high quality. First, we downloaded a list of the most cited papers from High Cited Papers of the Research Group in Engineering (on the Top 1 %) [9], and chose top 50 articles. Second, for each of 50 articles we took authors' keywords from a paper, identified less frequent ones, and removed it from the search query. Third, we ran a search in Web of Science with adjusted keywords and additionally limited this search by the year previous to the publication of the chosen article. The search resulted in articles with cited references used to create a graph of the area. Fourth, we created a citation graph using Web of Science for data (list of papers with cited references) and Sci2 for parsing and creating the citation graph [16].

There are different metrics of centrality to analyze the citation graph: Closeness Centrality [15], Graph Centrality [7], stress centrality [17], betweenness centrality [1, 6]. High centrality scores of a node show that it can reach others on relatively short paths, or that a node lies on considerable fractions of the shortest paths connecting others. Closeness Centrality is the most spread one [14] and calculated by the formula:

$$Cc(v) = \frac{1}{\sum_{t \in V} d_G(v,t)}, \quad (1)$$

where $d_G(v,t)$ is the length of the shortest path between the nodes v and t. If a network is not strongly connected, this measurement considered only reachable nodes. In some implementations, the centrality value for each node v is normalized by multiplying it by $n - 1$, where n is the number of nodes reachable from v.

We also decided to analyze InDegree as it provides information about citations of a paper in the selected context.

The graph centrality measure is the Page Rank algorithm that can provide interesting information about authoritative papers [3]. This algorithm is used as part of the Google search engine and ranks nodes in the graph by their importance.

And last but not least, this algorithm developed after PageRank by Kleinberg Hyperlinked-Induced Topic Search (HITS) is somewhat similar to PageRank, but makes a distinction between authorities and hubs [12]. The hubs are articles with numerous outgoing links, for example, reviews, and authorities are papers that have numerous incoming links from hubs.

Thus, we wanted to compare the findings of the citation graph analysis from contributions found using the keywords with InDegree, Betweenness Centrality (undirected graph), Closeness Centrality (undirected graph), PageRank, Authorities HITS and Hubs HITS with the actual reference list of the paper. All the calculations of centrality metrics were done in Python program using the NetworkX library (http://networkx.github.io/).

2.1 Data Description

For our analysis, we started with 50 cases from list of High Cited Papers of the Research Group in Engineering (on the Top 1%) [9]. To make a dataset, we followed a simple heuristics that the most precise keyword of the key phrase, which characterizes a paper, is the most rare one. To obtain better coverage of the area, we removed the less frequent keyword of the key phrase (if a search by the keywords presented in the article contains only one publication). For example, in the article A 2-tuple Fuzzy Linguistic Representation Model for Computing with Words by F. Herrera, L. Martínez [8] we find the following keyword phrases: information fusion, computing with words, linguistic variables, linguistic modeling. If we look at Google Scholar to examine frequency of the term, we will see that there are 2,940,000 articles with "information fusion", 2,680,000 with "computing with words", 1,670,000 with "linguistic variables", and 395,000 with "linguistic modeling" (as of the research time). The search by all the keywords provided us with 3 articles, including the paper from the list of top cited contributions, but two new articles were published in this field since 2000, so we could not use them in our analysis. If we remove the two rarest key phrases, we will receive 26 papers, all published since 2000, and the removal of three key phrases provides us with a list including more than 24,000 papers unsuitable for analysis. Adding the keyword "word" limited the results by linguistics area and allowed us to obtain 25 papers.

In our study we use this approach to adjust keywords and obtain better coverage of the research area. For each of 50 cases we downloaded search results from Web of Science, each paper coming with citation references (we only used data from Web of Science, as MEDLINE provides no information about citations references). The downloaded results were parsed in Sci2 to create a citation graph in the Graphml format. We then reversed and analyzed it using the devised Python program (where all the centrality coefficients were calculated using the NetworkX library). It should be mentioned that search queries were limited to the period before the case paper was published.

3 Analysis

We start our analysis with the most cited article A 2-tuple Fuzzy Linguistic Representation Model for Computing with Words by F. Herrera, L. Martínez [8], containing 449 citations. To create a citation graph, we made use of the adjusted keywords and limited the years of publication by 1999, as the article was published in 2000. The search result contains 25 papers, their bibliographical description with cited references downloaded from Thomson Reuters Web of Knowledge Database. Application of the Sci2 tool and the graph reversion using the Python script allowed us to develop a directed citation graph with 502 nodes and 533 edges.

Comparison showed that none of the papers with top 10 PageRank, Authority, Hubs, In Degree, Out Degree, Closeness Centrality, and Betweenness Centrality were presented in the actual reference list of the paper. We increased the number of top papers on the recommendation list to 20, but the result remained the same. Even if we look for top 30 papers with the highest coefficients they are not actually presented in the reference list of the first paper. Despite the obtained results that are disappointing, we continue analyzing the selected cases.

The outcomes of comparison recommendations and the actual list of references showed that in 18 of 50 cases the graph analysis failed to find even one paper cited in an actual paper. We created four sets of recommendations for reference list: top

Table 1 Average accuracy of recommendation versus centrality metrics

Amount of cases in recommendation list	7	10	20	30
PageRank	0.228	0.206	0.161	0.134
Authority (HITS)	0.171	0.157	0.122	0.097
Hub (HITS)	0.04	0.038	0.032	0.035
InDegree	0.245	0.22	0.174	0.153
OutDegree	0.037	0.038	0.033	0.036
Closeness Centrality	0.142	0.132	0.114	0.097
Betweenness Centrality	0.094	0.104	0.101	0.093

7, top 10, top 20, and top 30 papers received the highest values. Recommendation lists were created for all discussed centrality metrics. Comparison of the recommendation sets with actual literature showed that in 18 cases they were completely different. In 32 nontrivial cases the most accurate results were shown by InDegree, followed by those demonstrated by PageRank (see Table 1). It is worth mentioning that in all 7 cases where PageRank demonstrated better results it outperformed InDegree. InDegree outperformed all the other centrality metrics independent of how many papers (7, 10, 20, 30) we include in the recommendation list.

To see if PageRank provides additional information to InDegree, we compared the adjusted top10 recommendation lists (in which we only included actually cited papers). Analysis showed that in 5 cases PageRank found one additional paper, not included in the InDegree top 10 adjusted list, and in 1 case two additional papers.

4 Discussion

Although we predicted that centrality measures different from frequency of citation could provide better information, the analysis showed that InDegree delivered more accurate results. First of all, it supports the preferential attachment Barabási–Albert model [2] in the citation graph and could be an argument against using other measures. Notwithstanding the highly matching results obtained with PageRank and InDegree (which confirm the findings by Ma and this coauthors), we found PageRank capable of furnishing additional information, thereby enabling the development of complex recommendation metrics combining the values provided by both metrics.

Interestingly, in 18 cases we found that the centrality algorithms provided no information and all the centrality metrics failed to predict what sources would be used in the reference list of the paper with these keywords. There are three possible explanations for that:

First, the most cited papers usually discover a new research area and in this case it is hard to predict their reference list. Keeping in mind that we aim to create an automatic recommendation system for a scientific advisor, we may need different solutions for students and for experience authors. A solution for students could be developed by testing centrality metrics on papers published in the established area rather than on highly cited papers, which will allow us to use the same approach comparing the recommendation lists not with the case papers list, but with the actual list of the newest papers in this field.

The second explanation could be connected with keywords used for graph generation, as a small difference in keywords can lead to a major difference in search findings. There may be some mismatch among the keywords used by the authors of a paper and it is the keywords that actually characterize the paper. One of the interesting ways to continue our research is to use for search keywords automatically extracted from an article using TF-IDF or other metrics, rather than authors' keywords, to see if it could improve forecast quality.

The third explanation is related to our dataset limitations. In our dataset we use only sources available in Web of Knowledge and if we could develop a paper citation graph on data from Google Scholar, Scopus or other resource, we would obtain different results. Our dataset was developed for top cited papers in the Engendering field, but satiation could probably differ in another field. We will continue our research and testing similarity metrics on datasets from different fields. While it is evident that graph analysis can provide valuable information, it is worth mentioning that our analysis is only based on 50 cases, which restricts the generalization of the findings, and further analysis of a large sample of cases is needed. We see our research as the first step in this area, which will allow us to develop a methodology and look for ways to organize further research.

5 Conclusion

In our paper we selected the top 50 papers from the engineering field and run analysis to find out if any centrality measure can provide information about their actual reference list. We found InDegree to deliver the best results that can be still improved by adding information from PageRank. Analysis of the cases helped us formulate several new hypotheses about a possibility of finding breakthrough articles, the influence of keywords on the quality of recommendations, the role of the field and preferable citation patterns. In our further research we plan to increase the number of cases, analyze cases from different fields, and add automatic keyword generation.

References

1. Anthonisse, J.M.: The rush in a directed graph. Stichting Mathematisch Centrum. Mathematische Besliskunde (BN 9/71), 1–10 (1971). URL http://www.narcis.nl/publication/RecordID/oai:cwi.nl:9791

2. Barabási, A.L., Albert, R.: Emergence of scaling in random networks. Science **286**(5439), 509–512 (1999). URL http://www.sciencemag.org/content/286/5439/509.short
3. Brin, S., Page, L.: The anatomy of a large-scale hypertextual web search engine. Comput. Networks ISDN Syst. **30**(1), 107–117 (1998). URL http://www.sciencedirect.com/science/article/pii/S016975529800110X
4. Chen, C.: CiteSpace II: detecting and visualizing emerging trends and transient patterns in scientific literature. J. Am. Soc. Inform. Sci. Tech. **57**(3), 359–377 (2006). DOI 10.1002/asi. 20317. URL http://doi.wiley.com/10.1002/asi.20317
5. Chen, M.H., Huang, S.T., Hsieh, H.T., Kao, T.H., Chang, J.S.: FLOW: a first-language-oriented writing assistant system. In: Proceedings of the ACL 2012 System Demonstrations, ACL '12, p. 157–162. Association for Computational Linguistics, Stroudsburg, PA, USA (2012). URL http://dl.acm.org/citation.cfm?id=2390470.2390497
6. Freeman, L.C.: Centrality in social networks conceptual clarification. Soc. Networks **1**(3), 215–239 (1979). URL http://www.sciencedirect.com/science/article/pii/0378873378900217
7. Hage, P., Harary, F.: Eccentricity and centrality in networks. Soc. Networks **17**(1), 57–63 (1995). URL http://www.sciencedirect.com/science/article/pii/0378873394002489
8. Herrera, F., Martínez, L.: A 2-tuple Fuzzy Linguistic Representation Model for Computing with Words. IEEE Transactions on Fuzzy Systems **8**(6), 746–752 (2000). doi:10.1109/91. 890332.
9. Journal papers with more than 55 citations at web of science (last update: 21/10/2013) (2013). URL http://sci2s.ugr.es/highlyCitedPapers/index.php#top1
10. Joseph, M.T., Radev, D.R.: Citation analysis, centrality, and the ACL anthology. Ann. Arbor **1001**, 48,109–1092 (2007). URL http://clair.si.umich.edu/~radev/papers/112.pdf
11. Kinnunen, T., Leisma, H., Machunik, M., Kakkonen, T., Lebrun, J.L.: SWAN - scientific writing AssistaNt: a tool for helping scholars to write reader-friendly manuscripts. In: Proceedings of the Demonstrations at the 13th Conference of the European Chapter of the Association for Computational Linguistics, EACL '12, p. 20–24. Association for Computational Linguistics, Stroudsburg, PA, USA (2012). URL http://dl.acm.org/citation.cfm?id=2380921.2380926
12. Kleinberg, J.M.: Authoritative sources in a hyperlinked environment. J. ACM (JACM) **46**(5), 604–632 (1999). URL http://dl.acm.org/citation.cfm?id=324140
13. Ma, N., Guan, J., Zhao, Y.: Bringing PageRank to the citation analysis. Inform. Proc. Manag. **44**(2), 800–810 (2008). DOI 10.1016/j.ipm.2007.06.006. URL http://linkinghub.elsevier.com/retrieve/pii/S0306457307001203
14. Mihalcea, R.: Graph-based ranking algorithms for sentence extraction, applied to text summarization. In: Proceedings of the ACL 2004 on Interactive poster and demonstration sessions, p. 20 (2004). URL http://dl.acm.org/citation.cfm?id=1219064
15. Sabidussi, G.: The centrality index of a graph. Psychometrika **31**(4), 581–603 (1966). URL http://www.springerlink.com/index/u57264845r413784.pdf
16. Sci2 team: Science of science (sci2) tool (2009). URL http://sci2.cns.iu.edu.
17. Shimbel, A.: Structural parameters of communication networks. Bull. Math. Biophys. **15**(4), 501–507 (1953). URL http://link.springer.com/article/10.1007/BF02476438

Bounding Fronts in Multi-Objective Combinatorial Optimization with Application to Aesthetic Drawing of Business Process Diagrams

Julius Žilinskas and Antanas Žilinskas

Abstract The main concept of branch and bound is to detect subsets of feasible solutions which cannot contain optimal solutions. In multi-objective optimization a bounding front is used—a set of bounding vectors in the objective space dominating all possible objective vectors corresponding to the subset of feasible solutions. The subset cannot contain Pareto optimal (efficient) solutions if each bounding vector in the bounding front corresponding to this subset is dominated by at least one already known decision vector. The simplest bounding front corresponds to a single ideal vector composed of lower bounds for each objective function. However, the bounding fronts with multiple bounding vectors may be tighter and therefore their use may discard more subsets of feasible solutions. In this chapter we investigate the use of bounding vectors and bounding fronts in multi-objective optimization aided to aesthetic drawing of special graphs—business process diagrams. An experimental investigation shows that the use of the bounding front considerably reduces the number of function evaluations and computational time.

J. Žilinskas (✉)
Recognition Processes Department, Institute of Mathematics and Informatics, Vilnius University, Akademijos 4, LT-08663 Vilnius, Lithuania
e-mail: julius.zilinskas@mii.vu.lt

A. Žilinskas
Recognition Processes Department, Institute of Mathematics and Informatics, Vilnius University, Akademijos 4, LT-08663 Vilnius, Lithuania

Department of Applied Informatics, Institute of Mathematics and Informatics, Vilnius University, Akademijos 4, LT-08663 Vilnius, Lithuania
e-mail: antanas.zilinskas@mii.vu.lt

1 Introduction

A multi-objective optimization problem with d objectives $f_1(\mathbf{x}), f_2(\mathbf{x}), \ldots, f_d(\mathbf{x})$ is to minimize the objective vector $\mathbf{f}(\mathbf{x}) = (f_1(\mathbf{x}), f_2(\mathbf{x}), \ldots, f_d(\mathbf{x}))$:

$$\min_{\mathbf{x} \in X} \mathbf{f}(\mathbf{x}),$$

where $\mathbf{x}$ is the decision vector and X is the search space (the set of all feasible solutions). In most cases it is impossible to minimize all objectives at the same time, so there is no single optimal solution to a given multi-objective optimization problem.

Two decision vectors $\mathbf{a}$ and $\mathbf{b}$ from the search space can be related to each other in a couple of ways: either one dominates the other or none of them is dominated by the other. The decision vector $\mathbf{a}$ *dominates* the decision vector $\mathbf{b}$ (we denote $\mathbf{a} \succ \mathbf{b}$) if:

$$\forall i \in \{1, 2, \ldots, d\}: f_i(\mathbf{a}) \leq f_i(\mathbf{b}) \ \& \ \exists j \in \{1, 2, \ldots, d\}: f_j(\mathbf{a}) < f_j(\mathbf{b}).$$

The decision vector $\mathbf{a}$ is called a dominator of the decision vector $\mathbf{b}$ in this case. Decision vectors which are non-dominated by any other decision vector from the search space are called Pareto optimal or efficient and a set of these vectors is called a Pareto set or an efficient set. The set of corresponding objective vectors is called a Pareto front. Determination of these sets is the main goal of multi-objective optimization.

To present the proposed method more expressively an applied problem is used as a model problem. It is related to the diagrammatic visualization of business processes. The visualization here means the drawing of graphs where the vertices represent the business process flow objects and the paths represent the sequence flow. For the business process management, it is important to use correct, compact, and aesthetically pleasing diagrams drawn in notation respective to the standards of Business Process Model Notation [5]. Aesthetic layout of a business process diagram (BPD) is important for better understanding of the considered business processes, see, e.g., in [2, 7–9]. A BPD consists of a number of elements, e.g., activities, events, and gateways. The elements of diagrams are drawn as shapes which are allocated in a pool divided by the (vertical) swimlanes according to a function or role. The problem of the aesthetic graph drawing can be formulated as a problem of multi-objective optimization to optimize criteria of aesthetics [4, 10]. In the present chapter we consider graphs the vertices of which are located at the nodes of a rectangular grid and the edges are composed of segments of horizontal and vertical lines. The shapes are attributed to the swimlanes in the description of a business process. The problem considered here is to find the mutual disposition of the swimlanes and to allocate the shapes to nodes inside of swimlanes. Subsequently another combinatorial multi-objective optimization problem is solved where the edges of the graph are sought which represent aesthetically pleasing drawings of paths between the shapes; some algorithms for the latter problem are described in [4].

2 Branch and Bound for Multi-Objective Optimization

The main concept of branch and bound is to detect sets of feasible solutions which cannot contain optimal solutions. The search process can be illustrated as a tree with branches corresponding to subsets of the search space. An iteration of the classical branch and bound algorithm processes a node in the search tree that represents an unexplored subset of feasible solutions. The iteration has three main components: selection of the subset to be processed, branching corresponding to subdivision of the subset, and bound calculation. In single objective optimization, the subset cannot contain optimal solutions and the branch of the search tree corresponding to the subset can be pruned, if the bound for the objective function over a subset is worse than a known function value.

In multi-objective optimization bounding front may be used—a set of bounding vectors in the objective space dominating all possible objective vectors corresponding to the subset of feasible solutions. The subset cannot contain efficient (Pareto optimal) solutions if each bounding vector $\mathbf{b} \in B$ in bounding front B is dominated by at least one already known decision vector $\mathbf{a}$ in the current approximation S of the efficient set:

$$\forall \mathbf{b} \in B \; \exists \mathbf{a} \in S : \; \forall i \in \{1, 2, \ldots, d\} : \; f_i(\mathbf{a}) \leq b_i \; \& \; \exists j \in \{1, 2, \ldots, d\} : \; f_j(\mathbf{a}) < b_j.$$

The simplest bounding front consists of a single ideal vector composed of lower bounds for each objective function.

Figure 1 illustrates a bounding front. Dashed large circle circumscribes objective vectors (some are shown by dots) of a subset of the feasible solutions. Bounding vectors are shown with the largest solid circles shown in red and their dominated space is shown by the solid line. They bound the objectives over the subset of the feasible solutions: all objective vectors (dots) are dominated by (are higher than) bounding front (solid lines). Small circles represent currently known non-dominated objective vectors—they represent the current approximation of the Pareto front

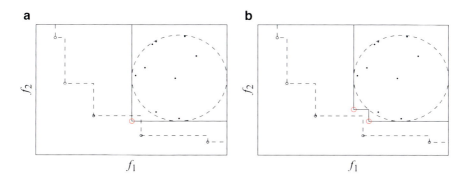

Fig. 1 Bounding front: (**a**) single bounding vector, (**b**) multiple bounding vectors

shown by dashed lines. These currently known non-dominated objective vectors do not dominate the single bounding vector shown in Fig. 1a, but both bounding vectors shown in Fig. 1b are dominated—they are higher than the dashed line. Therefore in such a situation branch and bound algorithm could discard the subset of the feasible solutions if bounding front with multiple bounding vectors is used, but cannot do so if the single bounding vector is used.

3 Multi-Objective Optimization Aided Allocation of Shapes in Aesthetic Drawings of Business Process Diagrams

We use allocation of vertices in aesthetic drawings of business process diagrams as an example multi-objective optimization problem in this chapter. BPDs consist of elements (e.g., activities, events, and gateways) which should be drawn according to the rules of Business Process Modeling Notation. The diagrams are drawn as shapes which are allocated in a pool divided by the vertical swimlanes according to a function or role. Here we are interested only in the allocation of shapes, i.e. we ignore the interpretation of the diagram in terms of the visualized business process.

According to the general opinion, the aesthetic attractiveness of the drawing of a BPD is especially important since the aesthetic layouts are also most informative and practical [1]. The graph drawing aesthetics is comprehensively discussed, e.g. in [2, 7, 8]. Although the problem of graph drawing attracts many researchers, and plenty of publications are available, special cases of that problem frequently cannot be solved by straightforward application of the known methods and algorithms. In the present chapter we consider a particular problem of drawing diagrams of business processes of small and medium size enterprises. The research is aimed at the development of the algorithm supposed for including into a relatively simple and not expensive software package discussed in more detail in [4] where also a complementary algorithms for the drawing connectors (edges of the considered graphs) are considered. The peculiarity of our approach is in explicit reduction of the problem of graph drawing into combinatorial multi-objective optimization problem.

A bi-objective optimization problem was formulated in [10] where the objectives are the length of connectors and the compatibility of the sequence flows with the favorable top–down, left–right direction. An algorithm based on the branch-and-bound approach was proposed with a single ideal bounding vector composed of lower bounds for each objective function formed by considering only already assigned shapes in a partial solution. In this chapter we investigate potential impact of shapes to be assigned later in the search on objective functions and building of a bounding front instead of a single bounding vector. An experimental investigation reveals usefulness of using the bounding front.

In this problem it is requested to allocate shapes in swimlanes, and the swimlanes with regard to each other aiming at aesthetical appeal of the drawing. The shapes are allocated in such a way so that the connected shapes were close to each other and that the connections would direct from left to right and from top to bottom. Two objectives are simultaneously optimized [10]:

- Minimization of the total length of connectors: The sum of city block distances between connected shapes is minimized.
- Minimization of the number of right down flow violations: The number of times the preceding shape in the connection is not higher than and is to the right from the following shape is minimized.

The shapes are allocated in a grid of predefined number of rows (n_r) and columns (n_c). The data of the problem are the roles (or functions), the shapes belong to, and the list of connections. Let us denote the number of shapes by n and the roles corresponding to shapes by $\mathbf{d}$, where d_i, $i = 1,\ldots,n$ define the role number of each shape. The connections are defined by $n_k \times 2$ matrix $\mathbf{K}$ whose rows define connecting shapes where k_{i1} precedes k_{i2}.

The shapes belonging to the same role should be shown in the same column (swimlane), however the columns may be permuted. Therefore part of decision variables define assignment of roles to columns. Let us denote the assignment of roles to columns by $\mathbf{y}$ which is a permutation of $(1,\ldots,n_c)$ and y_i defines the column number of the ith role. Another part of decision variables define assignment of shapes to rows. Let us denote this assignment by $\mathbf{x}$, where x_i defines the row number of the ith shape.

We define the objectives as following. The length of orthogonal connector cannot be shorter than the city block distance between the connected points. Therefore, we model the potential length of a connector as the city block distance between shapes. The total length of connectors is calculated as

$$f_1(\mathbf{x}, \mathbf{y}) = \sum_{i=1}^{n_k} |x_{k_{i1}} - x_{k_{i2}}| + |y_{d_{k_{i1}}} - y_{d_{k_{i2}}}|.$$

The number of right down flow violations is calculated as

$$f_2(\mathbf{x}, \mathbf{y}) = \sum_{i=1}^{n_k} v_d(k_{i1}, k_{i2}) + v_r(k_{i1}, k_{i2}),$$

where down flow violation is

$$v_d(i, j) = \begin{cases} 1, & \text{if } x_i \geq x_j, \\ 0, & \text{otherwise,} \end{cases}$$

and right flow violation is

$$v_r(i, j) = \begin{cases} 1, & \text{if } y_{d_i} > y_{d_j}, \\ 0, & \text{otherwise.} \end{cases}$$

The connection of two shapes in the same row violates down flow since the bottom or side of preceding shape connects to the top of the following shape.

In such a definition the objective functions are separable into two parts, one is dependent only on the decision variables $\mathbf{x}$ and another on $\mathbf{y}$:

$$f_1(\mathbf{x}, \mathbf{y}) = f_{1x}(\mathbf{x}) + f_{1y}(\mathbf{y}),$$

$$f_{1x}(\mathbf{x}) = \sum_{i=1}^{n_k} |x_{k_{i1}} - x_{k_{i2}}|,$$

$$f_{1y}(\mathbf{y}) = \sum_{i=1}^{n_k} |y_{d_{k_{i1}}} - y_{d_{k_{i2}}}|,$$

$$f_2(\mathbf{x}, \mathbf{y}) = f_{2x}(\mathbf{x}) + f_{2y}(\mathbf{y}),$$

$$f_{2x}(\mathbf{x}) = \sum_{i=1}^{n_k} v_d(k_{i1}, k_{i2}),$$

$$f_{2y}(\mathbf{y}) = \sum_{i=1}^{n_k} v_r(k_{i1}, k_{i2}).$$

Therefore the problem can be decomposed into two: find the non-dominated vectors (f_{1x}, f_{2x}) and the non-dominated vectors (f_{1y}, f_{2y}). The non-dominated solutions of two problems are then aggregated and the non-dominated solutions of the whole problem are retained. The number of solutions of the second problem is much smaller than that of the first problem. We approach the first problem by using a branch and bound algorithm. The second problem may be solved by a similar algorithm or even using enumeration of all solutions.

4 Branch and Bound for Assignment of Shapes to Rows in Business Process Diagrams

It was shown in the previous section that the optimization problem of allocation of shapes in aesthetic drawings of business process diagrams may be decomposed in two separate problems of the assignment of shapes to rows and the assignment of roles to columns. Let us discuss a branch and bound algorithm for the assignment of shapes to rows.

The assignment of shapes to rows is denoted by $\mathbf{x}$, where x_i defines the row number of the ith shape. Two shapes of the same role cannot be assigned to the same row:

$$x_i \neq x_j, \ i \neq j, \ d_i = d_j,$$

where d_i defines the role number of the ith shape.

Therefore the optimization problem is

$$\min \mathbf{f}(\mathbf{x}),$$

$$f_1(\mathbf{x}) = \sum_{i=1}^{n_k} |x_{k_{i1}} - x_{k_{i2}}|, \tag{1}$$

$$f_2(\mathbf{x}) = \sum_{i=1}^{n_k} v_d(k_{i1}, k_{i2}), \tag{2}$$

$$\text{s.t. } x_i \neq x_j, \ i \neq j, \ d_i = d_j, \tag{3}$$

$$x_i \in \{1, \ldots, n_r\}.$$

A set of solutions (subset of feasible solutions) may correspond to a partial solution where only some shapes are assigned to rows. Therefore, the partial solution is represented by the assignment $\mathbf{x}$ of $n' < n$ shapes to rows. The bounds for the objective functions include the direct contribution from the partial solution and the most favorable contribution from completing the partial solution [3]. Let us denote a bounding vector for the objective functions as

$$\mathbf{b}(\mathbf{x}, n') = \left(\sum_{i=1}^{n_k} c_1(i, \mathbf{x}, n'), \sum_{i=1}^{n_k} c_2(i, \mathbf{x}, n') \right),$$

where $c_1(i, \mathbf{x}, n')$ and $c_2(i, \mathbf{x}, n')$ denote the contribution of the ith connector to the bounds. When both connecting shapes are assigned in the partial solution, the direct contribution of the connector can be computed. In the contrary case, the most favorable contribution may be estimated. The bounds proposed in [10] involve only the direct contribution:

$$c_1^1(i, \mathbf{x}, n') = \begin{cases} |x_{k_{i1}} - x_{k_{i2}}|, & \text{if } k_{i1} \leq n', \ k_{i2} \leq n', \\ 0, & \text{otherwise}, \end{cases}$$

$$c_2^1(i, \mathbf{x}, n') = \begin{cases} 1, & \text{if } k_{i1} \leq n', \ k_{i2} \leq n', \ x_{k_{i1}} \geq x_{k_{i2}}, \\ 0, & \text{otherwise}. \end{cases}$$

A single ideal vector may be used as a bounding vector

$$\mathbf{b}^1(\mathbf{x}, n') = \left(b_1^1(\mathbf{x}, n'), b_2^1(\mathbf{x}, n') \right) \tag{4}$$

composed of two lower bounds for each objective function:

$$b_1^1(\mathbf{x}, n') = \sum_{i=1}^{n_k} c_1^1(i, \mathbf{x}, n'),$$

$$b_2^1(\mathbf{x}, n') = \sum_{i=1}^{n_k} c_2^1(i, \mathbf{x}, n').$$

Let us investigate a possible contribution of later assigned shapes in this chapter. If connected shapes belong to the same role, the vertical distance between them cannot be zero because two shapes cannot be assigned to the same row in the same column, see (3). Therefore, connectors contribute at least one to the vertical distance if they connect two shapes of the same role.

$$c_1^2(i, \mathbf{x}, n') = \begin{cases} |x_{k_{i1}} - x_{k_{i2}}|, & \text{if } k_{i1} \le n', \, k_{i2} \le n', \\ 1, & \text{if } k_{i1} > n' \text{ or } k_{i2} > n', \, d_{k_{i1}} = d_{k_{i2}}, \\ 0, & \text{otherwise.} \end{cases}$$

In such a case the bounding vector may be

$$\mathbf{b}^2(\mathbf{x}, n') = \left(b_1^2(\mathbf{x}, n'), b_2^1(\mathbf{x}, n')\right), \tag{5}$$

where

$$b_1^2(\mathbf{x}, n') = \sum_{i=1}^{n_k} c_1^2(i, \mathbf{x}, n').$$

If one of the connected shapes is already assigned, a favorable contribution of the connector may be estimated by looking at available places for the other connected shape:

$$c_1^3(i, \mathbf{x}, n') = \begin{cases} |x_{k_{i1}} - x_{k_{i2}}|, & \text{if } k_{i1} \le n', \, k_{i2} \le n', \\ \min_{x \ne x_j, \, d_j = d_{k_{i2}}} |x_{k_{i1}} - x|, & \text{if } k_{i1} \le n', \, k_{i2} > n', \\ \min_{x \ne x_j, \, d_j = d_{k_{i1}}} |x - x_{k_{i2}}|, & \text{if } k_{i1} > n', \, k_{i2} \le n', \\ 1, & \text{if } k_{i1} > n' \text{ and } k_{i2} > n', \, d_{k_{i1}} = d_{k_{i2}}, \\ 0, & \text{otherwise,} \end{cases}$$

$$c_2^3(i, \mathbf{x}, n') = \begin{cases} 1, & \text{if } k_{i1} \le n', \, k_{i2} \le n', \, x_{k_{i1}} \ge x_{k_{i2}}, \\ 1, & \text{if } k_{i1} \le n', \, k_{i2} > n', \, \nexists x > x_{k_{i1}} : x \ne x_j, \, d_j = d_{k_{i2}}, \\ 1, & \text{if } k_{i1} > n', \, k_{i2} \le n', \, \nexists x < x_{k_{i2}} : x \ne x_j, \, d_j = d_{k_{i1}}, \\ 0, & \text{otherwise.} \end{cases}$$

The bounding vector involving these contributions

$$\mathbf{b}^3(\mathbf{x}, n') = \left(b_1^3(\mathbf{x}, n'), b_2^3(\mathbf{x}, n')\right), \tag{6}$$

Bounding Fronts in Multi-Objective Combinatorial Optimization

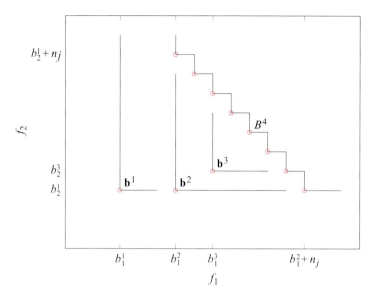

Fig. 2 Bounding vectors and bounding front

where

$$b_1^3(\mathbf{x}, n') = \sum_{i=1}^{n_k} c_1^3(i, \mathbf{x}, n'),$$

$$b_2^3(\mathbf{x}, n') = \sum_{i=1}^{n_k} c_2^3(i, \mathbf{x}, n'),$$

is better than one defined in (5), but it is more expensive to compute.

The most favorable contribution of the connector to the vertical distance is zero. It is achieved when the connected shapes belong to different roles and could be assigned to the same row. However in such a situation the contribution of the connector to down flow violation should be one because bottom or side of one shape connects to the top of the other shape. On the contrary, the most favorable contribution of the connector to down flow violation is zero when preceding shape is higher than the following shape ($x_{k_{i1}} < x_{k_{i2}}$). However in such a situation the vertical distance between the shapes would be at least one. Taking this into account a bounding front may be built:

$$B^4(\mathbf{x}, n') = \{(b_1^2(\mathbf{x}, n') + j, b_2^1(\mathbf{x}, n') + n_j - j) : j = 0, \ldots, n_j\}, \quad (7)$$

where n_j is the number of connectors where at least one of the shapes is not assigned in the partial solution and the shapes belong to different roles:

$$n_j = \left| \{i : k_{i1} > n' \text{ or } k_{i2} > n', \ d_{k_{i1}} \neq d_{k_{i2}}, \ i = 1, \ldots, n_k \} \right|.$$

Bounding vectors $\mathbf{b}^1(\mathbf{x}, n')$, $\mathbf{b}^2(\mathbf{x}, n')$, $\mathbf{b}^3(\mathbf{x}, n')$, and bounding front $B^4(\mathbf{x}, n')$ are illustrated in Fig. 2. The bounding front B^4 is better (tighter) than the bounding vectors, $\mathbf{b}^3$ is better than $\mathbf{b}^2$ which is better than $\mathbf{b}^1$.

The branch and bound algorithm for assignment of shapes to rows in business process diagrams may be built similarly to [10, 12] and can be outlined in short as follows:

1. Form the first possible assignment of shapes to rows in $\mathbf{x}$. Set $n' \leftarrow n + 1$.
2. Repeat while $n' > 0$

- **Evaluation of complete solution**: If the current solution is complete $(n' \geq n)$
  - Set $n' \leftarrow n$.
  - Compute

$$f_1(\mathbf{x}) = \sum_{i=1}^{n_k} |x_{k_{i1}} - x_{k_{i2}}|,$$

$$f_2(\mathbf{x}) = \sum_{i=1}^{n_k} v_d(k_{i1}, k_{i2}).$$

  - If no solutions in the current approximation S of the efficient set dominate the current solution $\mathbf{x}$, add it to the approximation:

  If $\nexists \mathbf{a} \in S : \mathbf{a} \succ \mathbf{x}$, then $S \leftarrow S \cup \{\mathbf{x}\}$.

  - If there are solutions in the current approximation S of the efficient set dominated by the current solution, remove them from the approximation:

  $S \leftarrow S \setminus \{\mathbf{a} \in S : \mathbf{x} \succ \mathbf{a}\}$.

- Otherwise
  - **Bounding**: Compute $\mathbf{b}^1(\mathbf{x}, n')$ using (4), $\mathbf{b}^2(\mathbf{x}, n')$ using (5), $\mathbf{b}^3(\mathbf{x}, n')$ using (6), or $B^4(\mathbf{x}, n')$ using (7).
  - **Pruning**: If $\mathbf{b}^1(\mathbf{x}, n')$, $\mathbf{b}^2(\mathbf{x}, n')$, $\mathbf{b}^3(\mathbf{x}, n')$, or every $\mathbf{b} \in B^4(\mathbf{x}, n')$ is dominated by a solution from the current approximation S of the efficient set, reduce n'.

- **Branching or retracting** (depth first search): Update $x_{n'}$ by available number and increase n' or reduce n' if there are no further numbers available.

Depth first search is used in the algorithm. The number of evaluated candidate sets (partial solutions in our case) may be a bit larger for this search strategy, but it saves memory and requires considerably less time for maintaining data structures [6]. This search strategy allows avoidance of storing of candidate sets [11], the number of the candidates may be very large in such problems when other search strategies are used.

5 Experimental Investigation

In this section we investigate performance of the algorithm for assignment of shapes in business process diagrams with various bounding vectors and the bounding front described in the previous section.

We perform computational experiments on problems of business process diagrams given in [10]. A computer with Intel i7-2600 CPU 3.40GHz, 8GB RAM, and Ubuntu 12.10 Linux was used for experiments. The branch and bound algorithm has been implemented in C/C++ and built with g++ 4.7.2 compiler.

The results of the experimental investigation are shown in Table 1. The computational time (t) and the number of functions evaluations (NFE) for the branch and bound algorithm are presented in Table 1. It can be seen that the use of bounding vector $\mathbf{b}^2(\mathbf{x}, n')$ (5) considerably improves the number of function evaluations and time compared to $\mathbf{b}^1(\mathbf{x}, n')$ (4). The use of bounding vector $\mathbf{b}^3(\mathbf{x}, n')$ (6) improves the number of function evaluations a bit further, but increases computational time since it is more expensive to compute. The use of bounding front $B^4(\mathbf{x}, n')$ (7) considerably improves the number of function evaluations and computational time. Using the bounding front reduces computational time up to almost 100 times compared to using the bounding vector proposed previously.

6 Conclusions

Branch and bound approach is well suitable for solution of multi-objective combinatorial optimization problems where the Pareto front can be reasonably bounded taking into account the direct contribution from a partial solution and the most favorable contribution from completing the partial solution to the objective functions. The proposed method is applicable for solution of multi-objective optimization problems related to the aesthetical drawing of business process diagrams.

Acknowledgements The support by Agency for Science, Innovation and Technology (MITA) trough the grant Nr.31V-145 is acknowledged.

Table 1 Results of the experimental investigation

n_r	$\mathbf{b}^1(\mathbf{x},n')$, (4) [10] t, s	NFE	$\mathbf{b}^2(\mathbf{x},n')$, (5) t, s	NFE	$\mathbf{b}^3(\mathbf{x},n')$, (6) t, s	NFE	$\mathcal{B}^4(\mathbf{x},n')$ t, s	NFE
Example problem								
6	0.15	237,440	0.06	930,469	0.08	704,048	0.01	187,343
7	1.44	28,118,029	0.33	6,074,083	0.49	4,576,302	0.04	656,290
8	9.49	192,605,603	1.54	29,855,682	2.26	23,101,727	0.18	2,593,238
Middle size problem								
5	0.87	11,846,524	0.14	2,292,133	0.19	2,110,564	0.04	518,681
6	8.76	87,341,601	0.86	15,097,449	1.26	14,111,040	0.21	2,993,714
7	20.10	267,553,983	2.21	40,710,474	3.40	38,251,546	0.52	7,370,189
8	37.05	473,246,383	3.41	64,644,742	5.37	60,846,181	0.83	11,886,008
9	76.96	997,982,630	6.72	128,330,033	10.66	120,741,102	1.31	18,437,102
10	193.69	1,946,020,628	13.17	257,442,963	21.22	243,423,005	2.23	47,220,762
11	394.98	3,386,280,514	25.03	487,597,206	39.77	464,519,182	8.50	131,752,014
12	751.75	5,496,804,470	46.13	949,050,115	76.33	914,075,489	24.45	397,440,621
13	1175.44	8,072,969,995	58.66	1,201,936,218	97.00	1,145,782,878	15.70	236,090,687
14	1845.78	11,516,056,991	85.80	1,774,663,616	143.27	1,695,153,806	22.48	353,554,807
15	2746.00	15,764,528,221	120.29	2,493,528,143	204.61	2,385,705,518	32.12	498,906,138
16	3825.00	20,848,903,023	161.08	3,363,454,730	270.26	3,222,389,040	44.81	672,931,502
17	5182.00	26,788,986,132	209.02	4,388,173,880	352.65	4,208,888,470	57.97	876,392,519
18	6817.00	33,597,007,137	263.56	5,570,100,374	445.95	5,347,708,471	73.53	1,109,646,700
19	8670.00	41,280,000,441	330.56	6,912,015,181	526.19	6,272,785,312	92.77	1,373,682,420

References

1. Battista, G.D., Eades, P., Tamassia, R., Tollis, I.G.: Graph Drawing: Algorithms for the Visualization of Graphs. Prentice Hall, Englewood Cliffs (1999)
2. Bennett, C., Ryall, J., Spalteholz, L., Gooch, A.: The aesthetics of graph visualization. In: Cunningham, D.W., Meyer, G., Neumann, L. (eds.) Computational Aesthetics in Graphics, Visualization, and Imaging, pp. 1–8. Elsevier/Morgan Kaufmann, San Francisco (2007)
3. Brusco, M.J., Stahl, S.: Branch-and-Bound Applications in Combinatorial Data Analysis. Springer, New York (2005)
4. Jančauskas, V., Mackutė-Varoneckienė, A., Varoneckas, A., Žilinskas, A.: On the multi-objective optimization aided drawing of connectors for graphs related to business process management. Comm. Comput. Inform. Sci. **319**, 87–100 (2012)
5. Owen, M., Jog, R.: BPMN and business process management. http://www.bpmn.org (2003)
6. Paulavičius, R., Žilinskas, J., Grothey, A.: Investigation of selection strategies in branch and bound algorithm with simplicial partitions and combination of Lipschitz bounds. Optim. Lett. **4**(2), 173–183 (2010). doi: 10.1007/s11590-009-0156-3
7. Purchase, H.: Metrics for graph drawing aesthetics. J. Visual Lang. Comput. **13**(5), 501–516 (2002)
8. Purchase, H., McGill, M., Colpoys, L., Carrington, D.: Graph drawing aesthetics and the comprehension of UML class diagrams: an empirical study. In: Proceedings of the 2001 Asia-Pacific Symposium on Information Visualisation, vol. 9, pp. 129–137 (2001)
9. Tamassia, R., Battista, G., Batini, C.: Automatic graph drawing and readability of diagrams. IEEE Trans. Syst. Man Cybern. **18**(1), 61–79 (1989)
10. Varoneckas, A., Žilinskas, A., Žilinskas, J.: Multi-objective optimization aided to allocation of vertices in aesthetic drawings of special graphs. Nonlinear Anal. Model. Control **18**(4), 476–492 (2013)
11. Žilinskas, A., Žilinskas, J.: Branch and bound algorithm for multidimensional scaling with city-block metric. J. Global Optim. **43**(2), 357–372 (2009). doi: 10.1007/s10898-008-9306-x
12. Žilinskas, J., Goldengorin, B., Pardalos, P.M.: Pareto-optimal front of cell formation problem in group technology. J. Global Optim. (2014, in press). Doi: 10.1007/s10898-014-0154-6